Culture and Technology

Culture and Technology

David S. Bertolotti, Jr.

Bowling Green State University Popular Press
Bowling Green, Ohio 43403

Library of Congress Catalogue Card No.: 84-71555

ISBN: 0-87972-307-6 Clothbound
0-87972-308-4 Paperback

Contents

Introduction

As part of my appointment as Professor of Humanities at GMI Engineering & Management Institute I was made responsible for a series of lectures in a course entitled "Society and Technology." Initially the course concerned itself with the impact of technology on society, and in 1969 such a stance seemed logical and correct. Indeed, one seemed constantly exposed to the fact from the lastest technological development, and, even more significantly, one became aware of the abundance of media coverage, both in print and electronics, which was devoted to existent and emerging technologies.

However, analysis and investigation into the relationship of technology and society led to the deterioration of the premise that technological impact constituted the major force in determining societal directions. In substance, it became obvious that there was a real, if not parallel, impact of society on technology. In other words, it appeared there was a constant give and take between society and technology, with no one force being either absolute or omnipotent, and the degree of interface between the two was yet unclear.

It also became apparent that the term "society" is woefully inadequate in defining the counterforce to technology. A more comprehensive concept is operative, something which incorporates the social, the psychological, the ethical, the metaphysical and yet numerous other facets of the amorphous "society." "Culture" seems, at least for the purpose of this book, to be more appropriate, for cultural likes and dislikes, cultural beliefs, cultural personalities and cultural ethics all combine into a temperament which seemed to constitute a powerful force in technological development. In other words, it is logical to assume that technologies exist because a large segment of any specific culture so wish those technologies to exist. Of course the computer is extant because of technical devices such as microchips, but as important is the desire on the part of some regiment of the population for computers. The essential issue, of course, is not how technology becomes part of the cultural consciousness but that a technology *is* part of the consciousness. One need not blunt moral or ethical considerations because such considerations are indeed important. However, the fact remains that a technology eventually weaves its way into a cultural consciousness, whether through need, advertising or as in the case of solar heating from a sub-culture impetus. Along with the technology is the intellectual climate and the emotional tone which give rise to that technology. In concise terms, technology not only impacts society, culture impacts technology.

With that premise somewhat firmly in mind I then had to find evidence to support such a contention. The major difficulty lies not so much in finding enough of the culture-technology interface but in finding examples which were specific enough to portray a definite philosophical stance somewhat isolated and clear. After careful deliberation on a great number of possibilities I decided on four: the Panama Canal, the aviation technology of World War II, the atomic bombing of Hiroshima and the development of the Salk anti-polio vaccine.

I chose these because each met criteria which I had established for acceptance. Each had to be from a different historical or philosophic era. Obviously the building of the Panama Canal is isolated both in thought and implementation from the others. It reflects a strong belief that the machines, however crude from modern standards, could accomplish part of the philosophic and political end of Manifest Destiny. World War II aviation development and the atomic bombing of Hiroshima are close in time historically yet different in philosophic relationships, roughly akin to a difference between the mechanical world (i.e., aviation as machines) and the Atomic Age (i.e., nuclear physics). In fact, the atomic event is seen by many as the beginning of a new era—true, either in a negative or a positive sense, yet a beginning. The fourth topic for investigation—the development of the Salk anti-polio vaccine—is reflective of a historical period overpowered with the wonders of chemistry. In a sense, the vaccine was researched and distributed in what up until that time constituted the most favorable time for the development of chemical compounds, cures and wonders. "Better Living Through Chemistry" was not only an advertising slogan, it was a cultural belief deeply rooted in a significant portion of the post-World War II population; and, as important, strongly subscribed to by those whose power lie in wealth, education and politics. In a very important way, the same kinds of forces which gave shape to Gothic cathedrals of earlier centuries also gave impetus to the explosion of science, specifically in regard to chemistry, in the 1950s. Centuries provide the examples, yet the process of translating culture into tangible respresentational objects is precisely the same. Those who controlled the likes and dislikes of the 12th century, those who held the religious and the economic balance of power caused magnificent cathedrals to rise; and, those who controlled the economic and technological balance in the 1950s ultimately contributed to the success of the Salk vaccine.

The next step in preparing the research required a determination of what adequately reflects the relationship between culture and technology. Particularly important was the consideration of communications. The news media, particularly news magazines and newspapers, provided ample discussion and deliberation on all four of the topics destined for amplification. The Panama Canal issue, for instance, was hotly debated in the press and in periodicals for decades. News of aviation technology, particularly that of German V-weapons, was included in daily and monthly reports from overseas. The stunning effect of the Hiroshima bombing filled the newspapers and magazines in August 1945. And the Salk vaccine was touted, supported, criticized, rejected, debated, evaluated and otherwise discussed as early as 1951 until it proved to be a satisfactory means of controlling polio.

In other words, the existing media spanned all four technological changes, thereby providing a common basis of comparison. Radio came too late for the Panama debate, and television late for all but the polio vaccine—and even then television was still in its infancy. However, a more important consideration lead me to choose the printed media: it best reflected the deep-rooted cultural attributes of each era. Part of this results from those in control of the press, those editors and publishers who, for good or evil, decided what was to be printed when. Also, the newspapers by their very nature give to slower and, in some cases, more careful deliberation of those issues. After all, the printed media reflected journalism in a more traditional sense than does either radio or the visual media. The very lack of rapidity was in effect a guarantee of an opportunity for a more comprehensive presentation of the news, allowing, of course, for inevitable inaccuracies and conscious propagandizing.

Finally, I attempted to create what novelist Henry James knew as "the tone of the moment." More accurately I recreate, in part, the tone of each era by highlighting cultural, economic and moral incidents from each period. International trade and nationalistic fervor essentially made the Panama Canal a necessity, albeit a non-profitable endeavor economically. In a somewhat objective manner, I have chosen those trappings which reflect some of the concerns, events, likes and dislikes occurring at the same time as the Panama-Nicaragua debate. With aviation technology, the reactions of elements of the populace are included, along with some of the favorite movies or social events. How one addresses the problem of wartime censorhip and propaganda is paramount; however, if one assumes the posture of the time, the propaganda was a fact of life and censorship a condition of communication. Who, after all, could not read between the lines? And, in a sense, there are those who expected and welcomed propaganda for immediate strategic reasons. Truth could, and to some degree would, be served in post-war peace, at least so claimed the confidently optimistic.

The reactions to the Hiroshima bombing, and even the events and attitudes preceding it, are indeed representative of, on the one hand, anxiety and concern before the bombing on how the war with Japan would end to relief following the bombing. However, the relief was tempered with ethical questions and political questions in the press. Reactions in the press to the Salk program as a whole reflected unbridled support and unabashed criticism; however, the general tenor of commentary is constantly optimistic—in keeping with the temper of the early 1950s.

In conclusion, the areas chosen for investigation provide unique opportunities for relaying the relationship of technology to culture, with the printed media of each era providing critical meaning to material which would otherwise constitute sterile technological incidents.

Chapter I

The Panama Canal: "The World's Greatest Engineering Feat"

Unlike the sudden intrusion of the atomic bomb and the relatively rapid employment of aviation technology, the planning and the building of the Panama Canal consumed decades. In fact if you were born on February 25, 1893—when an extensive argument raged as to whether the Panamanian or Nicaraguan route were better—you would be over ten years old before construction began in earnest. If nothing else, the canal project demonstrates quite adequately the role of politics, both internal and external, in technological change. Where it took less than a decade to put humans on the moon, it took over two decades for the United States government to simply maneuver around already established political realities related to the canal. In retrospect, that the canal was built at all is testimony that the technopolitical elite decide on what technologies are achieved; and at the turn of the century, the technopolitical concept of Manifest Destiny defined progress.

1: 1893-1903—Dreams and Schemes

The planning for canal construction took place at a time which essentially began to mark the end of agrarian America. The movement toward industrialization had begun and though not of a particularly sophisticated nature by later standards, the technology was beginning to make its presence known. In the midst of such movement, a political battle raged over the best canal route, and, predictably, many special interests were vocal in supporting their favorite. F.R. Brainard, writing in the *Scientific American* in 1893, claimed that "the present outlook for the completion of the Nicaraguan Canal is far better than for the Panama."[1] Work had commenced on the Nicaraguan route in 1878 and seemed to the United States Navy, for whom Brainard was writing, to be a more desirable route; besides, he added, the French plan had failed in Panama because of "inadequate knowledge of the country, incomplete surveys, wrong estimates and ruinous sub-contracts, and last, but not least corruption."[2] He was quick to add, however, that the company which would succeed would be the first one to receive the necessary funds.[3]

Others saw the canal as the inevitable end to four centuries of westward imperialistic movement: "For more than four centuries the mind of man has been possessed with a great idea."[4] International conditions would themselves be altered by the canal since apparent indifference would be replaced by American acceptance of the "attendant burdens" related to being "entangled in the affairs of the great family of nations."[5]

Some argued that economic profits would determine the desirability of canal ownership. The Suez Canal, for example, had cost $115,000,000 to build. In 1893 alone its profit was $8,333,333 on 3,559 vessels. Stock whose par value was $100 had climbed to $538.50, making the $20,000,000 worth owned by the British government valued at $95,000,000 on the open market.[6] However, continued the same report, the Nicaraguan Canal would be of far greater advantage since "the fate of the . . . Canal is still in doubt."[7] Also, according to the same analysis, the Nicaraguan Canal could only succeed, for even if it were a commercial failure it "would be of great advantage to the United States, as controlling ownership would be worth $70,000,000."[8]

There were those, however, who saw the canal project for what it was: a thinly veiled attempt at international one-upmanship. The English had Suez, the French had failed in Panama, and the Germans were casting their eyes toward the Western Hemisphere with unbridled expansionistic desires. Why, indeed, should the United States not have a canal of its own? *The Nation*, in "The Canal Mania," argued that the canal was not desirable, from any justifiable standpoint:

> It is hard to name any considerable canal except Suez which is profitable. It is unnecessary to refer to that awful loss of life and property at Panama. The folly displayed there was too stupendous to be repeated.[9]

But the project was not to be long put down by such argument. Readers were far more impressed by such statistics as the 50,000,000 cubic yards which the French had already excavated, leaving *only* 65,000,000 cubic yards to be moved—all for only $100,000,000.[10] In a world where a brass bed could be bought for $1.98 and "Fancy worsted flannels" and "tweed men's trousers" for $3.50, the average citizen must have been overpowered by the rhetoric which encompassed hundreds of millions of dollars (figures 1,2,3). One can somewhat imagine a parallel to that of a farm peasant entering a Gothic cathedral.

Who indeed can comprehend such magnificence? And who indeed can challenge those who create such extraordinary assertions?

The public was assured, however, that those in control had control: "The Panama Canal is on sound footing and is within calculable distance of completion."[11] Charts, diagrams and plans were incorporated to inform the public of statistical details in order to correct the "broad difference between the Panama Canal as it actually is and the Panama Canal as it exists in the Public's eye."[12] "Public ignorance" and "general misunderstandings" were addressed and "corrected," even to the point of claiming that mortality rates, health figures and climatic dangers had been exaggerated, especially since "Negroes from the British Antilles" could be used to counter any environmental hazards.[13] It seems almost anything could be justified in the name of Manifest Destiny:

> The desire of the American people to have a canal at the Isthmus built with their own money and operated under their own control is very natural and proper.[14]

As the argument continued, three schooners were "driven ashore by [a] gale and probably lost" off the Cape Cod coast, including the three-masted *Alice B. Phillips* (621 gross tons) and the three-masted *Helena*.[15] Col. Teddy Roosevelt's Rough Riders were to visit Chicago to lay a cornerstone for a new post office.[16] Typical of the exaggerated rhetoric of the time, the National League outlined a plan to fight the American Association: "By coming to our territory . . . the association invites war."[17] It seems some types of expansionism in the 1800s were more desirable than others.

Without delay, an American Isthmus Canal Company began on $30,000 (which promised an increase to $250,000,000) with these objectives: one, to build a canal from the Atlantic to the Pacific; two, to colonize, build, own and operate any canal from ocean to ocean with aid from the United States, England, France, Germany and/or any other country.[18] The company had not yet selected a route, but investors were reassured that a "sea-level canal" (code term for Nicaraguan) had backing from "capitalists of importance" who chose not to be identified.[19] However, the government was not about to let private investment alone dictate the course of the canal adventure. By July 11, 1899 an Isthmus Canal Commission was selected to investigate the possibility of

AUGUST 12, 1900.

RUINS OF PANAMA CANAL

$225,000,000 More Necessary to Complete Waterway's Construction.

MILES OF ROTTING MACHINERY

Only About One-third of the Work Done —Appalling Waste of Millions of Dollars.

Special to The New York Times.

LOS ANGELES, Aug. 11.—"The most woeful and reckless waste of wealth in these modern days is that of the costly machinery, enormous dredges, steamboats, locomotives, railroad tracks, and cars on the Isthmus of Panama," said ex-State Senator Leonard H. Perkins the other day.

The Senator has been in Colon and Panama for several weeks this season and he has had the opportunity to learn many things about the great projected canal of De Lesseps and the French investors across the isthmus. "Fully $50,000,000 that came out of the proverbial stockings of the French peasantry," he continued, "lies rotting and rusting in the mud and muck of the isthmus. It is appalling to see the prodigality, the shameful waste of French money everywhere apparent along the route of the proposed canal from Colon to Panama City.

"The work on the Panama Canal, under the management of the de Lesseps company, was abandoned in 1889, with the collapse of the company's fortunes. Five years of labor and some $135,000,000 had been actually spent upon the proposed canal, besides about $130,000,000 more which had been wasted. The great army of superintendents, sub-superintendents, engineers, machinists, contractors, and the negro laborers from Jamaica and Yucatan quit work as soon as the news of the fall of the de Lesseps company reached the Isth-

Fig. 1
New York *Times*, August 12, 1900, p. 3.

TRANSFER TO

Bloomingdales

3rd AVE 59th & 60th Sts

Manufacturers Stocks

"Sale of Manufacturers' Stocks."

Women's Waists.

Of white lawn, all over embroidery fronts, tucked back and flare cuffs; all sizes,

69c

Second Floor, 59th St. Section.

"Sale of Manufacturers' Stocks."

Handkerchiefs.

Half a dozen items, representing better values than it has ever been our privilege to offer.

Men's hemstitched and initialled pure linen, at $1.00 per half dozen, or, each 18

Men's hemstitched and initialled pure linen, at $1.45 per half dozen, or, each 25

Men's hemstitched and initialled cambric, at 75c. per half dozen, or, each 12½

Women's all linen initial at, per box of half dozen 50

Women's unlaundered initial, bowknot design, at 90c per half dozen, or, each 15

Women's hemstitched and lace trimmed, in two lots, at 6 for 25c and 6 for 50

Main Floor, 59th St. Section.

"Sale of Manufacturers' Stocks."

Laces and Embroideries.

Several hundred pieces 18 inch cream Cluny and Venise Allover Nets at half value, per yd., 39c., 29c., 15

Less than a hundred cartons of French Val. Laces, values up to 40c., at per dozen yds., 12c. to 25

1,500 yards Liberty Silk, 42 inches wide, light gray, brown, leghorn, salmon, lilac, nile, maise, navy, old rose, helio; regular value 75c. a yard; for to-morrow only, per yd., 19

Another shipment of 12,000 yds. embroideries, manufacturers' sample strips, insertings and edgings, slightly imperfect, but certainly a bargain at 25c. on the dollar; choose at, per yd., 5c. and 10

Main Floor, 59th St. Section

"Sale of Manufacturers' Stocks."

Pure Silk Umbrellas.

26 and 28 inch Men's and Women's pure silk twill Umbrellas, with case to match, handles in large variety, actual value $3.00 to $3.50, at

1.90

Main Floor, 59th St. Section.

"Sale of Manufacturers' Stocks."

Fig. 2

New York Times, August 12, 1900, p. 3.

Fig. 3
New York Times, August 1, 1900.

canal construction. Readers were informed that the engineers were "the very best in the country" and that many "already acquired useful knowledge of the special work at hand."[20] However, even supporters of the commission were forced to admit that except for riding "a railroad over the Panama route" little could be done during the rainy season.[21]

ii. The Hay-Pauncefote Treaty

Not surprising, enthusiasm ran high for a canal—almost *any* canal—early in 1900—for enthusiasm was the timbre of the times. Telephone service was available for residents in Manhattan for about sixty dollars a year. Electric lights were lit by gas, either from the New Amsterdam Gas Company at fifty cents per 1,000 cubic feet or from the Mutual Gas Light Company at sixty five cents per 1,000 cubic feet.[22] "Cornelius Vanderbilt's health has improved since his return to Newport from Europe."[23] In Bar Harbor, Maine, the Kebo Valley Clubhouse had burned down but was soon to be replaced.[24] And despite "a heavy fog at Southampton" which "prevented any display of pyrotechnics" the Fourth of July holiday was pleasant and successful.[25] Of course not all were caught up in the enthusiasm and excitement of the times. In one letter to the editor of the New York *Times* a reader complained about short-skirted bike riders:

> There can be no excuse for women mounting a man's wheel, and further, they should not be permitted to do so. If fathers, mothers, and brothers will not prohibit such exhibitions, the law should be powerful to intervene on behalf of decency.[26]

If decency applied to bicycle riding, it soon lost its way in the debate over acceptance of the Hay-Pauncefote Treaty. Even while the treaty was being negotiated, readers were exposed to invocations of the Monroe Doctrine and calls for an "All-American" canal. England was being ousted from participation even before the treaty advocating a co-custodial policy was drafted: "We shall be expected to keep it open and protect it as Great Britain protects the integrity of the Suez Canal by her control of Egypt."[27] Such early rumblings of isolationism did not deter supporters of the treaty from predicting that the treaty had virtually no opposition: "Senator Morgan promises there will be little difficulty in getting votes," "The Canal Bill Will Pass," "Not More Than Half a Dozen Votes Against It."[28] And as a sequel to the treaty, readers were told that the United States might just as well purchase the Danish West Indies as a refueling station.[29] However, readers were advised that such action in the Pacific Galapagos had better be delayed since "Ecuador has developed an aversion to surrendering the smallest speck of territory."[30] The Washington *Post*, at first a strong supporter of the treaty, began to call for rejection of the treaty. Claims were made that Secretary Hay had been "hypnotized by London society," but still the country "would take his high integrity for granted."[31] The paper added: "We do not believe, however, that intelligent public opinion will improve the connection."[32]

> Either we have the right to construct a canal across the Isthmus, or we have not. If we have not, no European nation can give it to us. If we have, there is no occasion to ask England's permission to exercise it.[33]

Who bound to the Monroe Doctrine and inspired by Manifest Destiny could argue against such logic? "Rejection of the treaty by the Senate would be evidence of astounding blindness and bad judgment,"[34] said one editorial. Supporters of the treaty argued that the treaty made the canal possible. Detractors argued that England was simply looking for a way to exert prestige in the Western hemisphere. Supporters claimed the opposition was attempting to discourage neutrality. Detractors claimed that England wanted the United States to absorb the cost of the canal and England would provide nothing in return (fig. 4). The battle was engaged.

> "Repudiate The Treaty"
>
> It would be for the general welfare of the civilized world, but not for the defense of the United States, since it would be open in time of war to our enemies on the same terms as us.[35]

The Washington *Post* added: "Let us have a canal by all means," but not with England.[36] The New York *Times* called for passage of the treaty since the world benefits from guaranteed neutrality and since the treaty has the "general approval" of the Central American nations. The Panama Canal Company urged delay until the return of the Isthmus Commission.[37] The *Post* claimed "a New York Syndicate" controlled the Nicaraguan Concessions, and two of the investors were identified: John Jacob Astor and George Westinghouse.[38] Those who support the treaty

Fig. 4

Detroit Evening News, Feb. 7, 1900, front page.

support shrewdness, "dark insinuations" and "secret understandings."[39] "Shall We Ratify A Blunder?" asked the *Post*, "To ratify the Hay-Pauncefote Treaty is to abandon forever the Monroe Doctrine."[40] President McKinley's leanings became a factor with many senators, so much so that headlines read "Ratification Is Predicted."[41] The New York *Times* saw no conflict with the Monroe Doctrine, and about the treaty states: "It will, we are confident, commend itself to the American people as soon as they turn from false teachers."[42] The Detroit *Evening News* probably expressed the most objective position concerning ratification, and later history shows that its assessment was accurate: "The information from Washington [is that] there was quite a number of senators not quite ready to swallow the Hay-Pauncefote Treaty."[43] Such positions proved satisfactory to the *Evening News* since they indicated "gratifying evidence that Americanism is not dead in that body" (fig. 5).

The *Evening News* readers also found they could buy a beautiful set of teeth for $4.98, that fillings were fifty cents and painless extractions were only a quarter. For those who cared more for candy than sound teeth, advertisements informed of Monday candy specials with chocolate covered, peppermint peanut clusters or coconut cream for fifteen cents a pound. A chorus of one hundred voices and a piano playing ragtime concern would be featured at the Last Cake Walk of the Season at Grand Central Palaces. In Cleveland, Ohio, in March of 1900 Mrs. Lily Langtry's appearance at the Euclid Avenue Opera House was cancelled since "Cleveland has a censor ready for 'The Degenerates'," who would cause the performance to be shut down if "anything objectable be said or done on the stage."[45]

However, Buffalo Bill's "Wild West Show and Congress of Rough Riders" featuring the "Charge Up San Juan Hill" went on as scheduled, as did the Metropolitan's farewell concert to John Phillip Sousa and his official American Band (for the Paris Exhibition). One could win a live lion at the 1900 Convention Carnival (fig. 6). *Ben Hur* was on Broadway, and Huber's 19th Street Museum featured such equine wonders as a three-legged Boer horse and a talking horse guaranteed to answer questions and solve mathematical problems.

The issue of the Hay-Pauncefote Treaty became less important with the introduction of two other canal factors: the rekindlings of the Nicaraguan-Panama Debate and the long awaited Canal Commission Report. Local news, such as pay for teachers, attracted much attention. Female teachers were paid

TREATY'S FATE

Hay-PauncefoteConvention May Be Amended

To Abolish Clayton-Bulwer Deal and Allow Fortifying.

The Document Seems to Have Stirred Up a Tempest.

(By Associated Press.)

WASHINGTON, Feb. 7.—The senate committee on foreign relations met today to consider the Nicaragua canal treaty. This discussion was animated. While some opposition was manifested it was not of such a character as to indicate the defeat of the treaty; but that it will be amended there seems to be little doubt.

One of the amendments proposed is that the whole of the Clayton-Bulwer treaty shall be abrogated. It is said

Fig. 5
Detroit Evening News, Feb. 7, 1900, front page

EMPIRE THEATER EXTENDED ENGAGEMENT.

This Afternoon 2:15; Tonight 8:15; Also Monday Afternoon and Evening.

OWING TO THE PHENOM-ENAL SUCCESS OF THE **GENUINE PICTURES** OF THE

PERFORMANCE Suitable for Ladies

JEFFRIES-SHARKEY FAMOUS FIGHT

These are the Pictures Taken by the American Biograph Company.

MATINEES DAILY. 25-50-75 EVENING PRICES. 25-50-75-$1

WIN A LION

$10,000 WORTH OF PRIZES

GIVEN AWAY.

1900 CONVENTION CARNIVAL

Old Wonderland Building,

30 WOODWARD AVE.

All This Week COMMENCING MONDAY

Afternoons 2 P. M. Evenings 8 P. M.

Continuous Vaudeville—Haunted Swing—Monte Carlo—Parisian Visions—Bijou Theater.

ADMISSION—25 CENTS.

Fig. 6

New York Times, March 22, 1900, entertainment section.

six hundred and sixty-four dollars a year, with forty dollars a year increase up to fifteen years; male teachers received more for the same work—nine hundred dollars a year, with the same increments. And one H.E. Gansworth proved to be an entrepreneurial wizard:

"An Indian Honored at Princeton"

H. E. Gansworth, a full blooded Tuscarawa Indian was appointed junior class orator by the faculty today. He supports himself by compiling abstracts of notes and lectures and by selling newspapers among students.[46]

Meanwhile the canal debates continued: "House Votes for Nicaragua Canal 225-35," an action which prompted one editor to suggest it was a "foolish thing to do" in advance of treaty ratification.[47] Nicaragua, of course, had already expressed an interest in the canal by offering to sell the needed land at a favorable price. The Isthmus Canal Commission assumed the position of political expediency by deciding it could not report until December of 1900, eight months after its return to the United States.[48] The sum of $1,000,000 was appropriated for the Walker Commission to investigate the canal proposition even more fully by employing 1,000 engineers: 500 in Nicaragua, 300 in Panama, and 200 along the Darien route.[49] It seemed in May of 1900 that the Nicaragua route was favored. Further support for Nicaragua was indicated in Senator Leonard E. Perkins' attack on Panama.

The most woeful and reckless waste of wealth in these modern days is that of costly machinery, enormous dredges, steamboats, locomotives, railroad tracks, and cars on the Isthmus of Panama.

No engineeer down here believes it would ever be finished.[50]

The $225,000,000 *more* needed for completion (fig. 7), the miles of rotting machinery, the appalling waste of millions of dollars, and the fifty million taken from "the stockings of French peasants" all lead Senator Perkins to state: "No, I have little faith in the Panama Canal."[51] Henry L. Abbot took issue with such appraisals. In the *Atlantic Monthly* he argued in an extensive comparisons of costs, of hours and minutes, of years (and meters) of dirt that Panama was the best route. He disagreed with Perkins by showing that two-fifths of the work had been already

AUGUST 12, 1900.

RUINS OF PANAMA CANAL

$225,000,000 More Necessary to Complete Waterway's Construction.

MILES OF ROTTING MACHINERY

Only About One-third of the Work Done —Appalling Waste of Millions of Dollars.

Special to The New York Times.

LOS ANGELES, Aug. 11.—"The most woeful and reckless waste of wealth in these modern days is that of the costly machinery, enormous dredges, steamboats, locomotives, railroad tracks, and cars on the Isthmus of Panama," said ex-State Senator Leonard E. Perkins the other day.

The Senator has been in Colon and Panama for several weeks this season and he has had the opportunity to learn many things about the great projected canal of De Lesseps and the French investors across the isthmus. "Fully $50,000,000 that came out of the proverbial stockings of the French peasantry," he continued, "lies rotting and rusting in the mud and muck of the isthmus. It is appalling to see the prodigality, the shameful waste of French money everywhere apparent along the route of the proposed canal from Colon to Panama City.

Fig. 7
New York *Times*, August 12, 1900, p. 4.

completed.[52] Abbot also suggested that to disagree was futile:

> It is certain that if we are to have the best possible canal to connect our coasts, it must cross the Continental divide at Panama. Nature has so decreed, and it is idle to contest the decision.[53]

But many did contest the decision, including the very powerful Isthmian Canal Report which called the Nicaragua route "the most practical and feasible route ... under management ownership of the United States" (fig. 8). The Commission's findings were sent to Congress, including the estimated cost of $200,540,000 and ten years to build.[55] Difficulties in securing rights and concessions were expected with Panama but not with Nicaragua. Also, the Panama Canal was more distant from major cities: 377 miles more to New York, 576 miles more to New Orleans, and 386 miles more to Liverpool.[56] The British comment on the Nicaraguan Canal was non-supportive: "It is a fascinating scheme, but we have some doubts if the canal would be worth cutting."[57] In the midst of such argument, the Hay-Pauncefote Treaty was gaining support. Early committee support had been 9-2 against, now it was 5-5 or 6-4 in favor.[58] Of course this was according to the New York *Times* estimate that claimed detractors would "insult the Secretary of State every day or two" and had a "hatred of England that pervades a certain class of votes."[59]

On March 14, 1901 Benjamin Harrison died at 4:45 o'clock in the afternoon. His death was reported as being both quiet and painless. The same could have been said for the prospects of the Hay-Pauncefote Treaty: "The outlook for a renewal of negotiations between America and Great Britain over the Nicaragua Canal is very gloomy."[60] Finally, on May 16, 1901, the New York *Times*, which had been the strong supporter of an American-English canal venture was forced to concede that "the present outlook is better for the building of the canal without any agreement."[61] On November 3, 1901 a decision was made to formally terminate joint-ownership, the instrument being a modified proposal:

> The treaty dissolves the partnership control over the canal. . . and gives the United States the right to build and defend the canal. It was drawn up in Washington and Great Britain has assented to it.[62]

This second Hay-Pauncefote Treaty was signed on

ISTHMIAN CANAL REPORT

The President Sends Commission's Findings to Congress.

NICARAGUA ROUTE APPROVED

Estimated Cost, $200,540,000—Would Accommodate Largest Vessels and Take Ten Years to Build.

WASHINGTON, Dec. 4.—The report of the Isthmian Canal Commission, submitted by the President to Congress to-day, gives as the unanimous conclusion of that body that "the most practicable and feasible route for an isthmian canal, under the control, management, and ownership of the United States, is that known as the Nicaragua route."

The commission estimates the cost of this route at $200,540,000. This estimate is much in excess of any heretofore made, and is due to increased dimensions and other features not heretofore considered.

Fig. 8

New York Times, Aug. 12, 1900, p. 4.

November 19, 1901 and declared by England as a one-sided bargain for the United States which claimed the signing was a literal acceptance of the Monroe Doctrine. However, readers of the Detroit *Evening News* were told that even the second treaty was unacceptable for "as the convention now stands Briton [sic] gains ... without expense to herself for construction maintenance, or operation."[63] As a response to the apparent American gain, *The Saturday Review* advocated a German-English friendship for "holding the U.S. in check."[64] Finally, all support for a joint venture was silenced by December 6, 1901 when the second treaty was ratified by a vote of 72-6. Hosteltor's

Stomach Bitters offered the only respite for Anglophiles, for one dose would "tone nerves, cure stomach disorders, safeguard against insomnia, and build firm flesh."[65]

At least one issue had been resolved after years of debate: the United States would build a canal without foreign intervention; however, no one had yet decided where the canal was to be. The stage was being prepared for a massive technological achievement, somewhere perhaps *anywhere*, even if it proved to be economically unprofitable.

iii. The Century Turns: From Chaos, Chaos

What was it like in 1901? Fashion-minded newspaper readers found they could purchase Alaska Sable Scarfs for $12, Sable Fox Boas for $9.50, Mink Scarfs for $15, twenty-four inch Seal coats for $250, and Grand Duke Business Suits from $15 to $35. English toe oxfords sold for $1.45, flour for $4.75 a barrel, ham for $.11 a pound and a leg of lamb for $.13 a pound. Andrew Carnegie was surprised that his $5,200,000 gift to New York City caused any sensation. Parasols with black wooden handles (in plain delicate colors) were stylish. There was concern that King Edward's new yacht had been incorrectly designed, thereby making it top heavy. Two non-union men were shot to death in Columbia, South Carolina, during the Southern Railroad Strike. Readers of the *Yonkers Statesman* slapped their knees over the following:

Yeast: "Would you call this automobile a runabout?

Crimsonbeak: "Yes. It will run about ten miles then break down."[66]

One prominent doctor connected with a hospital for the insane claimed that one-half of all foreign born residents were insane, based on the fact that 50 percent of all inmates were foreign born. His solution? Each immigrant must arrive on American shores with a "sanity certificate from an inspector stating the immigrant was sane."[67]

"Oil Discovered in Egypt"

A terrific gas flow followed the discovery and caused an explosion which wrecked the boring plant and blocked up the well.[68]

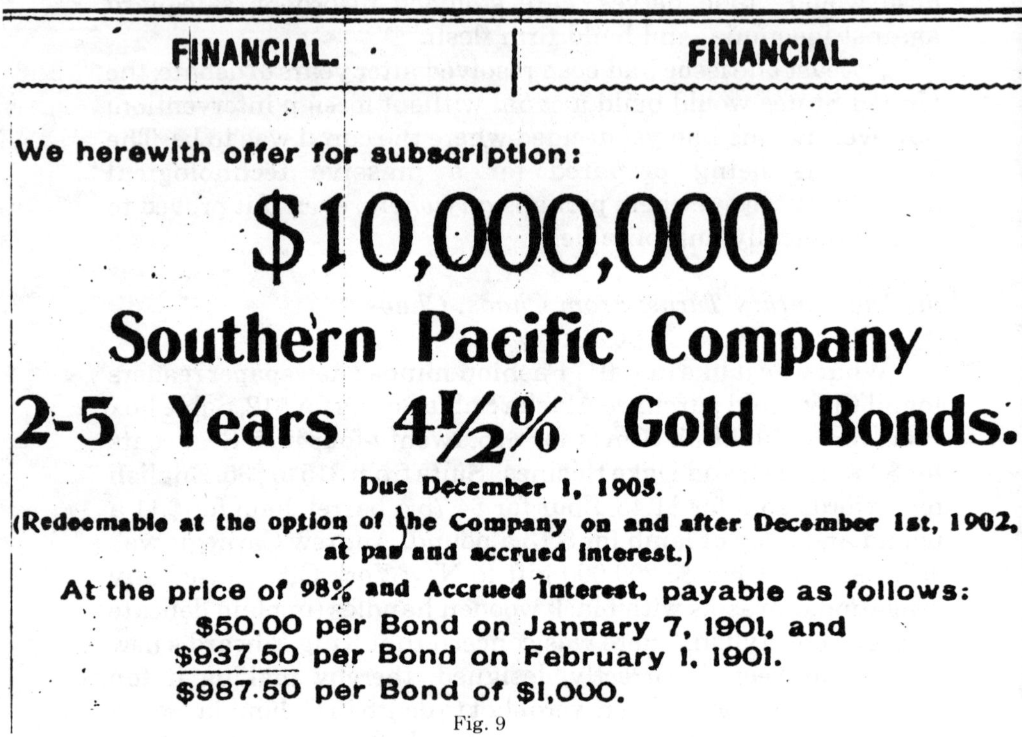

FINANCIAL. FINANCIAL.

We herewith offer for subscription:

$10,000,000

Southern Pacific Company

2-5 Years 4½% Gold Bonds.

Due December 1, 1905.

(Redeemable at the option of the Company on and after December 1st, 1902, at par and accrued interest.)

At the price of 98% and Accrued Interest, payable as follows:

$50.00 per Bond on January 7, 1901, and
$937.50 per Bond on February 1, 1901.
$987.50 per Bond of $1,000.

Fig. 9
New York Times, Aug. 12, 1900, p. 4.

Advertised investments were available everywhere (fig. 9). One could invest in oil. Or one could purchase a seat on the Stock Exchange for $75,000. A house on Fifth Avenue, between 64th and 65th Streets cost $500,000 (35′ by 100′). Or if one wanted sidelights, there was a house available on East Fifth Street for $325,000. Too high? One could become his own landlord in Queens for between $2,000 and $3,700. Edison phonographs were $1 down and $1 a week. The Russian government was guaranteeing four-percent bonds, payable in U.S. gold on a semi-annual non-taxable basis. The bonds were made tax free by Imperial Russian decree and a sinking fund of 1/2 of 1 percent was being raised annually to protect the loan. (Unfortunately for investors the bonds were scheduled to mature in 1957.) Or, one could invest in a new Isthmus Company which had acquired the rights for the construction of the Panama Canal, established for the purpose of "acquiring, owning, and operating all canals, harbors, railroads, anywhere in Central America, the United

States or Columbia, or South America."[69] $2000 capitalization promised growth to $1,500,000. One person who said he would not invest in the Panama Canal was Andrew Carnegie. When asked why, he answered:

> Decided not . . . I do not believe it would be workable if built, and as for the Panama route, that, if built, I do not believe it would render one-half the service that a double-track railroad would. Even if no tolls were charged it would be a losing proposition. There is more sentiment than common sense behind the canal question.[70]

And sentiment it was that indeed did build the canal! In addition to Manifest Destiny and military advantage, economic determinism supplied ample emotional stimulus for the canal existence:

> Our national prosperity and the increase of our national wealth, the opportunity for profitable employment of capital, the brains, skill, the muscle and energy of the American people depend on getting and keeping foreign markets for our surplus profits.[71]

iv. The Canal Commission's Non-choice

The critical issue *where* to build the canal had not yet been settled. Panama? Nicaragua? The Darien Route? Columbia? As Miss Alice Roosevelt was introduced to society in what most papers reported as the most charming event since the long past days of Dolly Madison, new battle lines were drawn. The French Panama Company was rumored ready to sell its concessions, yet details of the sale were not available. A selling price of $40,000,000 had been bandied about, but no action was taken.

Panama seemed to be the inevitable choice by March of 1901, especially with the suggestion by General Walker that the full commission report *may* recommend Panama. Both the United States and Columbia had held secret negotiations, with readers being informed that Columbia had stated rigid conditions of the Concession transfer to the United States. In addition, the French Concession had been extended to 1910, thereby keeping the French Panama Company in a viable trading position. In fact, there were many reports that the Columbian government would insist on the French selling the Concession. W. H. Edwards, head of the Nicaragua surveyors, emphasized the Nicaragua route was very practicable; however, because of an admitted lack of knowledge, he reserved judgment on Panama's potential.

Meanwhile, the equipment in Panama continued to rust since the political advantage had not yet been decided. One issue was finally quietly resolved—the signing of the second Hay-Pauncefote Treaty in November of 1901: "There was no particular ceremony connected with the signature yesterday of that important convention."[72] In effect, the treaty was no longer newsworthy. The New York *Times* conceded defeat of the Nicaragua route with the following comparison:[73]

	Nicaragua	Panama
1) Cost:	$189,864,062	$144,233,357 plus 109,141,000 concession prices) minus 40,000,000 done by French
2) Harbors:	none	many
3) Construction time:	6 years	10 years
4) Locks:	yes	sea level
5) Passage:	33 hours	12 hours
6) Projected Complications:	none local	many local

Based on this analysis, the following conclusion was drawn: "It would seem probable that our Government will now decide to give further consideration to the Panama project."[74] *Scientific American* reached a different conclusion on the same Commission Report, and published:

> . . . we certainly think that the attitude of the French owners has been such as to render the recommendation of the Nicaragua route the only logical course open to the commission.[75]

Claims and counter claims flowed. The Panama canal trip would be 134.6 miles shorter than the Nicaragua Canal. The French claimed the Nicaragua route was impossible. Britain apparently didn't expect any canal to be built; therefore, it happily signed the second Hay-Pauncefote Treaty. Supposedly, every member of the Commission supported the Panama route. Opponents to the Panama route were branded as opponents to *any* canal. Senator Morgan (Alabama) claimed it was a question of public morality, not engineering feasibility or cost, for according to him, the United States should not deal with men who had robbed the French people of their life savings.

More confusion resulted from the Columbian decision to issue permits only to private capitalists and not to foreign governments as stated in Article XXI and XXII of the Concession Agreement. Americans wanted a concession in perpetuity—not one for ninety-nine years; thereby another block to construction was presented. However, by April 1, 1902, Columbia was convinced to give her consent to the sale of Panama and to concede necessary building rights. The Panama Canal Company was to sell its rights for $40,000,000. Columbia was to get $1,000,000 for its holdings, and an annual rental of $750,000.[76]

However, when the treaty was signed on April 24, 1902, some significant changes were reported. One report claims Columbia was to get $7,000,000 and charge no rent for fourteen years. Another report claimed no price had been fixed for concessions.[77] Careful notice was given that a full treaty had been signed. It was not protocol "as has been mistakenly reported in the press."[78] Confusion reigned over what Columbia really wanted, as the technology and the engineers waited.

There was no confusion at Macy's department stores where no credit, discounts or commissions were given in order to eliminate direct expense and credit losses. The final score of the Harvard-Yale baseball game left no confusion—Harvard won 6-5. 'Little dusky skinned' Filipinos had "taken great fancy" to the national game of baseball, and were interested, too, in both football and track. The Philadelphia *Times* reported:

> Now, since Mark Twain has been made an L.L.D., it will be said, we suppose, that he will in the future lay down the law. Yet he never took it up.[79]

The Boston University College of Liberal Arts faculty recommended seventeen students for Phi Beta Kappa, fifteen of them women. Apparently, such action had little influence on the Board of Trustees at Northwestern University, which decided to abolish its female medical school:

> It is impossible to make a doctor out of a woman. Women cannot grasp the chemical and pharmaceutical laboratory work, the intricacies of surgery, or the minute work of dissection.[80]

It must have seemed to newspaper readers in 1902 that it too was impossible to build a canal in Central America. Senate debates, local disagreements, strikes, international intrigue, bribes, largesse and fear of malaria all provided seemingly impossible roadblocks to the "world's greatest engineering wonder."[81] The *Nation* assured its readers that whatever happens, the canal will be built—and Panama was the only choice:

There can be no doubt that the House will say ditto to the Senate, in the matter of the Isthmian Canal Bill, and that, accordingly, the Panama route will be adopted by the Congress of the United States.[82]

The article went on to state that "the *Evening Post* was the first and only newspaper of any consequence to maintain the Panama route was best.[83] At first, this particular newspaper was "hooted at" for its simple-mindedness; however, it proved to be right in a "triumph of reason over prejudice."[84] In addition, a favorable Inter-Technical Committee Report in 1898 had "a positive effect on engineers" so that it was possible to realize "the dream of a hundred years."[85] Reason over prejudice? The *Nation* went on to claim that the Nicaragua route was "a mania, a cult."[86] *Scientific American*, earlier in 1902, had reached a similar position by suggesting that from an engineering standpoint and from an historical standpoint the Panama route was best.[87] It claimed the Panama route was also best from "the standpoint of common sense."[88]

As the Spooner Bill authorized negotiation to establish the best route,[85] "Mrs. Carrie Nation [lectured] the Vanderbilts at a horse show on the evils of overdressing."[90] While the Columbian government was being described as a "vulgar striker" and the "holder-up of an important enterprise for selfish purposes of extortion,"[91] the following was happening in Indiana:

"Negro Who Assaulted Two Women Hanged to Telephone Pole"

A mob of forty to fifty took the prisoner from the officers. The negro was taken to the homes of both of his victims, and by them was identified . . . one mile east of the [John] Lemon farm, a rope was thrown over a telegraph pole and the negro was hanged. After hanging the negro the mob quietly dispersed.[92]

As the cry of "By Treaty or By War" engulfed Columbia's intransigency, the Chancellor of the University of Nebraska was attacking the evils of football by claiming the dismantling of a street car was a result of this "devil's work." The students who performed the evil deed had just attended a football game.[93]

v. Does Anyone Understand the Final Treaty?

While the amount of dollars to be spent on the building of the canal was being debated, there was no question over the destiny of 400,000 Finns who were starving, following a crop failure caused by spring frosts, a cold summer and great floods. It was reported as the worst famine since 1861 when 100,000 died.[94]

Rear Admiral Walker found himself defending the Canal Commission's expenditures of $1,036,481, of which $240,000 was applied to salaries and traveling expenses. He probably would have preferred going to West Palm Beach, Florida, where he could have stayed at The Palms for two or three dollars a day, with weekly rates available. Or, better yet, special winter rates were available at The Hotel Agnew in Atlantic City—$2.50 a day and up! Of course this included a separate department for the sexes and electric auto service to the rail station.

In February 1903 coal prices had been reduced from $7.50 a ton to $6.50 a ton. Irish linen handkerchiefs were eight cents each, corn whiskey cost $2.85 a gallon, and Armor and Swift were expected to lose $500,000 because of overbuying butter (bought at 17-18¢ per pound, forced to sell at 14-16¢ a pound). America had to be reassured that Germany was not after the Panama Canal, and Australia planned to stem immigration from the general depression in South Africa by requiring that each immigrant either have a "billet" to go to or have $100 in one's possession. The Guggenheim Exploration Company was investing $50,000,000 in Mexican mining, and it was reported that Standard Oil interests were also identified with the deal. An Anglo-German demonstration in Venezuelan waters was declared a clear violation of the Monroe Doctrine. A woman teacher was to stand trial for the offense of matrimony, since married women were declared not fit for the "arduous duties" of assistant superintendent. Germany again declares no interest in the Canal.

Readers were told the canal delay was nearly over and were given advice on how to protect women aliens "who are ignorant of the character of the real houses in which they have been promised employment."[95] Advertisements claimed "Woman's Safeguard" in Pennyroyal Pills or Wilcox Tansy Pills. Men could restore their manhood with "Cathos"—five day free trial (fig. 10). If that

Test for Yourself the Wonderful Curative Properties of Swamp-Root

To Prove What This Great Remedy Swamp-Root Will Do for YOU, Every Reader of the "News-Tribune" May Have a Sample Bottle Free.

Reporters Have Convincing Interviews With Prominent People in Greater New York Regarding Wonderful Cures.

(From New York Journal, January 4th, 1900.)

65TH POLICE PRECINCT,

DR. KILMER & CO., GREATER NEW YORK, NOV. 11, 1899.

How to Find Out If You Need Swamp Root.

It used to be considered that only urinary and bladder troubles were to be traced to the kidneys, but now modern science proves that nearly all diseases have their beginning in the disorder of these most important organs.

The kidneys filter and purify the blood—that is their work.

So when your kidneys are weak or out of order, you can understand how quickly your entire body is affected, and how every organ seems to fail to do its duty.

If you are sick or "feel badly," begin taking the famous new discovery Dr. Kilmer's Swamp-Root, because as soon as your kidneys are well, they will help all the other organs to health.

Disordered kidneys are responsible for more sickness and suffering than any other disease. [illegible] irritates the nerves, makes you [illegible] restless, sleepless and irritable. Makes you pass water often [illegible]

Fig. 10

New York *Times*, Feb. 11, 1900

failed, one could try Mormon Bishop's Pills, which not only provided Vim, Vigor, and Vitality but also cured one from self-abuse, dissipation and cigarette smoking. Odine's Tablets would not only restore vitality but also cure one from "Unfitness to Marry" and eliminate constipation. Readers were informed that Columbia needed to come to terms on the Panama Canal or the Nicaraguan route would again be considered.

The Treaty with Columbia was signed by Secretary Hay on January 16, 1903, yet the route was not officially decided:

> If we ever undertake to the building of a canal at Panama—it is too early to say that we shall undertake—we may find reason to regret bitterly our choice of that route.[96]

However, from this point on the canal question was quickly resolved. On March 4, 1903 the Panama Canal Company accepted the American offer. The Senate ratified the Panama Treaty (only 5 votes against) by March 18, 1903. On March 19, 1903, there was a rush of applicants for the Canal Commission. The French were satisfied, the Columbia Congress still strongly approved the treaty, and the London *Times* claimed the United States Senate had yielded to great presidential pressure. In April, plans were being made for General Walker to personally inspect the proposed route with Major William Black of the Sanitary Division of the Corps of Engineers (malaria again!). The engineers were ready, the technology poised for what was indeed to be "the greatest engineering feat." It was now simply a matter of moving everything into place, curbing disease, hiring workers and beginning to dig the Panama Canal.

On November 20, 1903, the *Nation* in a retrospective view of the treaty signing, said:

> Columbia's appeal . . . is at once dignified and pathetic. It is the voice of a weak people, asking considerate treatment by a nation of resistless power.[97]

About two weeks before the Senate ratified the treaty, Miss Alice Roosevelt went to the horse races. Sweet Alice (6-5) ran in the second race, but finished third. The race was won by Royal Deceiver, which paid 7-1.[98]

Notes

[1]Brainard, F.R. "The Panama and Nicaragua Canals," *Scientific American* (Vol.

LXVIII, No. 8), Feb. 25, 1893, p. 119.

[2]*Ibid.* [3]*Ibid.*

[4]Mahan, A.T., "The Isthmus and Sea Power," *Atlantic Monthly* (Vol. LXXII, No. CLCC XXXII) Oct. 1893, p. 459.

[5]*Ibid.*

[6]"Ship Canals Projected and in Progress," *Scientific American* (Vol. LXXII, No. 6) Feb. 9, 1895, p. 90.

[7]*Ibid.* [8]*Ibid.*

[9]"The Canal Mania," *The Nation* (Vol. LXI, No. 1596), Sept. 12, 1895, pp. 180-81.

[10]"Nicaragua Canal Debate: Project Again Under Discussion in the Senate," New York *Times*, Jan. 18, 1899, p. 5.

[11]"The New Panama Canal," *Scientific American* (Vol. LXXX, No. 5), Feb. 4, 1899, p. 73.

[12]*Ibid.* [13]*Ibid.*, p. 75.

[14]"An American Canal at Panama," *Scientific American* (Vol. LXXX, No. 10), March 11, 1899, p. 146.

[15]"Stranded on Cape Cod Coast," New York *Times*, May 4, 1899, p. 2.

[16]New York *Times*, May 20, 1899, p. 3.

[17]New York *Times*, Feb. 4, 1900, p. 21.

[18]"New Ship Canal Scheme," New York *Times*, June 27, 1899, p. 4.

[19]*Ibid.*

[20]"A Good Example To Follow," New York *Times*, July 11, 1899, p. 6.

[21]"Isthmusian Canal Commission," New York *Times*, editorial, p. 3.

[22]Apparently competition was a reality before the times of utility monopolies and multiquinical public utility commissions.

[23]"The Gossip and Doings of Peoples at Various Summer Resorts," New York *Times*, July 19, 1899.

[24]*Ibid.* [25]*Ibid.*

[26]"Women's Bicycle Dress," Letter to the Editor, New York *Times*, July 11, 1899, p. 6.

[27]"All American Canal," New York *Times*, editorial, p. 6.

[28]"The Canal Bill Will Pass," Washington *Post*, Feb. 3, 1900, p. 3.

[29]"Sequel to Canal Treaty," Washington *Post*, Feb. 8, 1900, p. 4.

[30]*Ibid.*

[31]"Repudiate the Treaty," Washington *Post*, Feb. 7, 1900, p. 3.

[32]*Ibid.* [33]*Ibid.*

[34]"Opposition to Hay-Pauncefote Treaty," New York *Times*, Feb. 8, 1900, editorial, p. 6.

[35]"Repudiate the Treaty," Washington *Post*, Feb. 8, 1900, p. 6.

[36]*Ibid.*

[37]New York *Times*, Feb. 9, 1900, p. 5.

[38]"Claims All Canal Rights," Washington *Post*, Feb. 10, 1900, front page.

[39]"Shall We Ratify a Blunder," Washington *Post*, Feb. 9, 1900, p. 6.

[40]*Ibid.* [41]*Ibid.*, p. 3.

[42]New York *Times*, Feb. 14, 1900, p. 6.

[43]"Treaty's Fate, *The Evening News* (Detroit), Feb. 7, 1900, front page.

[44]*Ibid.*

[45]"Mrs. Langtry's Date Cancelled," New York *Times*, March 22, 1900, p. 7.

[46]"An Indian Honored at Princeton," New York *Times*, March 22, 1900, p. 7.

[47]*Ibid.*

[48]"The Nicaragua Canal Debate," New York *Times*, May 2, 1900, p. 3.

[49]"Nicaragua Canal Debate," New York *Times* May 2, 1900, p. 3.

[50]"Ruins of the Panama Canal," New York *Times*, August 12, 1900.

[51]*Ibid.*

[52]Abbot, Henry L., "The Best Isthmian Canal," *Atlantic Monthly* (Vol. LXXXVI), Dec. 1900.

[53]*Ibid.*

[54]"Isthmian Canal Report," New York *Times*, Dec. 5, 1900, p. 7.

[55]*Ibid.* [56]*Ibid.* [57]*Ibid.* [58]*Ibid.*

[59]"Watch Them," New York *Times*, Dec. 5, 1900, editorial, p. 8.

[60]"Canal Negotiations May Not Be Resumed," New York *Times*, March 14, 1901, p. 2.

[61]New York *Times*, May 16, 1901, editorial, p. 8.

[62]"Lord Pauncefote Back With Canal Treaty," New York *Times*, Nov. 3, 1901, front page.

[63]"The New Treaty of No Great Value to U.S.," *The Evening News (Detroit)* Dec. 6, 1901, p. 2.

[64]*Ibid.*, Dec. 7, 1901, front page.

[65]Advertisement, New York *Times*, March 14, 1901, p. 2.

[66]"Naygets," New York *Times*, Nov. 20, 1901, p. 8.

[67]"Half-Foreign Born Insane," Washington *Post*, Dec. 6, 1901, p. 8.

[68]"Oil Discovered in Egypt," New York *Times*, Dec. 12, 1901, p. 3.

[69]"New Isthmus Company," New York *Times*, July 2, 1901, p. 5.

[70]"Andrew Carnegie Back," New York *Times*, Nov. 3, 1901, p. 7.

[71]"Successful Diplomacy," New York *Times*, Nov. 19, 1901, page 8.

[72]"New Treaty Signed," Washington *Post*,Nov. 19, 1901, front page.

[73]"The British Idea of A Treaty With Liberal Concessions," *The Evening News* Nov. 19, 1901, front page.

[74]"Isthmian Canal Report," New York *Times*, Dec. 5, 6, 1901.

[75]"Report on the Isthmian Canal Commission" (Vol. LXXXIII, No. 24), Dec. 15, 1901, p. 370.

[76]"Canal Protocol Given," New York *Times*, April 1, 1902, p. 3.

[77]"Panama Canal Treaty Signed," New York *Times*, April 24, 1902, p. 3.

[78]*Ibid.*

[79]"Mark Twain and the Law," *Nayget*,New York *Times*,June 8, 1902.

[80]"Says It Is Impossible To Make Woman a Doctor," New York *Times*, Jan. 5, 1902, p. 3.

[81]At best, there seemed no ingredients for the technological achievement of building the canal; one wonders what would have happened in an advanced technological society with instant medicines.

[82]"The Best Canal Assured," *The Nation* (Vol. 74), June 26, 1902, pp. 498-499.

[83]*Ibid., p. 498.* [84]*Ibid.* [85]*Ibid.* [86]*Ibid.*

[87]"Panama or Nicaragua—Which?" *Atlantic Monthly* (Vol. LXXXVI, No. 3), Jan. 18, 1902, p. 34.

[88]*Ibid.*

[89]"Turns to Nicaragua," Washington *Post*, Nov. 21, 1902, p. 4.

[90]*Ibid.*, front page.

[91]"Not by War," New York *Times*, Nov. 24, 1902, p. 8, editorial.

[92]"Lynched by Indians Mob," Washington *Post*, Nov. 21, 1902.

[93]"Football and the Devil," New York *Times*, Nov. 7, 1902, p. 8.

[94]"400,000 Finns Starving," New York *Times*, Dec. 20, 1902, front page.

[95]"Women to Protect Aliens," New York *Times*, Jan. 15, 1903, p. 3.

[96]"The Treaty With Columbia," New York *Times*, Jan. 26, 1903, p. 8.

[97]"The Appeal to Conscience," *Nation* (Vol. LXXVII, No. 2004), Nov. 20, 1903, p. 418.

[98]"Miss Roosevelt at the Races," New York *Times*, Feb. 22, 1903.

Chapter Two
The Second World War and Aviation Technology

Anything new in war, while you wait in a confined space, speculating over the nature of a mysterious horror, inordinately protracts the night.[1]
H.M. Tomlinson

There indeed must have been many protracted nights in World War II because of the extraordinarily large amount of technological development which took place; however, particularly fearful to both the Allies and the Axis Powers was the war from the air. Somehow it seemed to be more mysterious and less comprehensible to those beings whose existence was earthbound to a considerable extent. In a curious way, aviation technology created devices which were at the same time awesome and awful. The newspapers and periodicals which reported on aviation technology—and most did—were able to capture both the mystique and the mystery.

i. 1941 Inventions: Ingenious, Innocent or Insane

Not all aviation technology was of the magnitude of the V2, radar or the Norden bombsight.[2] In fact, the New York *Times* had a weekly feature entitled "Novel Patents of the Week" which included such inventions as an antiaircraft shell loaded with a grappling hook attached to a parachute designed to slow down an enemy plane with a sufficient amount of drag, allowing the plane to become an easy target for groundfire.[3] Or equally curious was another type of shell which was designed, upon explosion, to entangle enemy planes in a net of steel cables.[4] However such inventiveness was not unique to aviation materials—in fact, it seems the early years of the war, 1941 and 1942, viewed technological development much the same as it had been perceived early in the 20th Century. Discoveries were made by individuals creating implements to satisfy immediate ends or inquisitive tastes. A reflective cane for walking at night seems logical enough, however a cage for pinboys at bowling alleys, a woodpecker doorknob and neon Christmas tree lights seem at best frivolous. And the Swoose—a hybrid between a goose and a swan—seems somewhat unnecessary. Fortunately, nearly all offspring were sterile.[5]

As the war progressed, the inventions and the related technologies became more serious and more war related. Part of this obviously was because the war became an overpowering influence on virtually all the American people, and part was due to the restrictions put on information, material and manufacturing. The press not only reported that "Japanese Forces Wipe Out Western Luzon; Midway, Wake Resists";[6] and that "The Russians Continue to Advance on All Fronts,"[7] but also that "Cosmic Rays Tell Plane Altitude" because of a newly discovered relationship between altitude and cosmic ray intensity.[8] The same newspaper which carried an article telling how over a hundred hostages were shot in France by the Nazis also informed the American reader that the patent (No. 2,265,966) for the invention of the cosmic ray altimeter was being awarded to Rudolph Gebauer of Berlin, Germany.

In a way, the technological mentality seemed as naive and as confused as the American public itself, and the movies of 1941 reflected this uncertain state to a degree with a strange combination of domestic comedy and moderate propaganda dominating: Jimmy Durante and Phil Silvers in *You're In The Army Now*, Abbott and Costello in *Keep 'Em Flying* and Ed Sullivan's stage show with the "Dawn Patrol Review" competed with *Dumbo, The Birth of the Blues, Look Who's Laughing* (Fibber McGee and Molly), *How Green Was My Valley* and Bogart's *Maltese Falcon*. It was almost as though war was a minor nuisance not yet fully understood. In fact, one edition carried a photograph of the planet Venus below a crescent shaped moon with the caption: "By Way of Change: A Non-War Picture."[9]

The emergency rations of a Japanese plane shot down over Pearl Harbor on December 7, 1941 seemed newsworthy: a bottle of whiskey, chocolate impregnated with whiskey, calorie free candy, toothpaste, and chopsticks.[10] As the war heated up, however, so did the environment from which technologies develop (fig. 11). Patent No. 2,267,615 for retractable seaplane floats was issued (to Germany's Dorier Manufacturing Plant, however). Electrically heated pilot suits for high altitude flying and for aiding wounded airmen were being developed (a similar German suit was declared "practically worthless").[11] The Southeast Air Corps Training Center at Maxwell Field boasted it had never found a faultily packed parachute.[12] Beryllium was being used more and more in engine parts, and it was listed as a strategic material in November of 1941 because of a worldwide shortage which was caused by "Hitler storing up as much as he could lay his hands on,"[13] or so the readers were told. In fact, while other countries were collecting and storing gold, Germany was said to have been "very content to store up metal which could be made into war material."[14]

Folding aircraft wheels reflected a high stage of development (only the German Ju87-B was without them),[15] "radio trails" which gave accurate information as to the plane's location and altitude, a new invention which became known as the "turbo-jet," and dual automatic directional finders all portended the degree of aviation development yet to follow. The American press told its readers that planes would soon be equipped with television since radio was imperfect at best; that the new SXB2C-1 dive bomber could carry heavier loads at higher speeds for longer distances; that refinement of liquid oxygen containers would lighten a plane by three hundred pounds (thereby allowing for greater armament); that Flexseal—a new safety glass ten times stronger than auto safety glass—was being designed for substratospheric airplanes (fig. 12).

Yet, at best, America was unprepared for war, and the popular press, despite attempts to glamorize technological developments, reflected the confident confusion. Sonja Henie as "Queen of the Hollywood Ice Review" certainly captured more attention than the silica-gel which had been invented to protect engines from rust in transit. How could one ignore the opportunity to purchase an RCA Victrola for only $79.95 (with one year to pay). A trip to Trinidad at $160 for fourteen days or one to Miami at $56.95 for eleven days certainly had more appeal in 1941 than a muffler to silence airplane engines, even to the amazement of many who "wonder why the European War has not produced a satisfactory air muffler."[16] Cultured pearls at $9.98, lightweight English bikes at $34.95 and square Scotch decanters at $5.98 claimed equal newspaper space with the report that Australia had ordered a coastal blackout. The price of handmade Williamsburg glass (70¢) or metal cookie baskets ($1.39) was mixed in with advice on what to do in an air raid or stories of the increased firing power of planes to hundreds of rounds per minute (fig. 13). How to make your tires last longer, the O-O tie of the Army-Navy football game, and the fact that one could learn to tango and to rumba at Miss Julie's Dance School in three weeks probably drew much more attention than the following:

Fig. 11

Advertisements such as these inflamed war emotions. Home movie advertisements appear in *Vanity* and other publications.

Fig. 12

As in so many similar advertisements, the war technology was sold in a commercial-civilian mode.

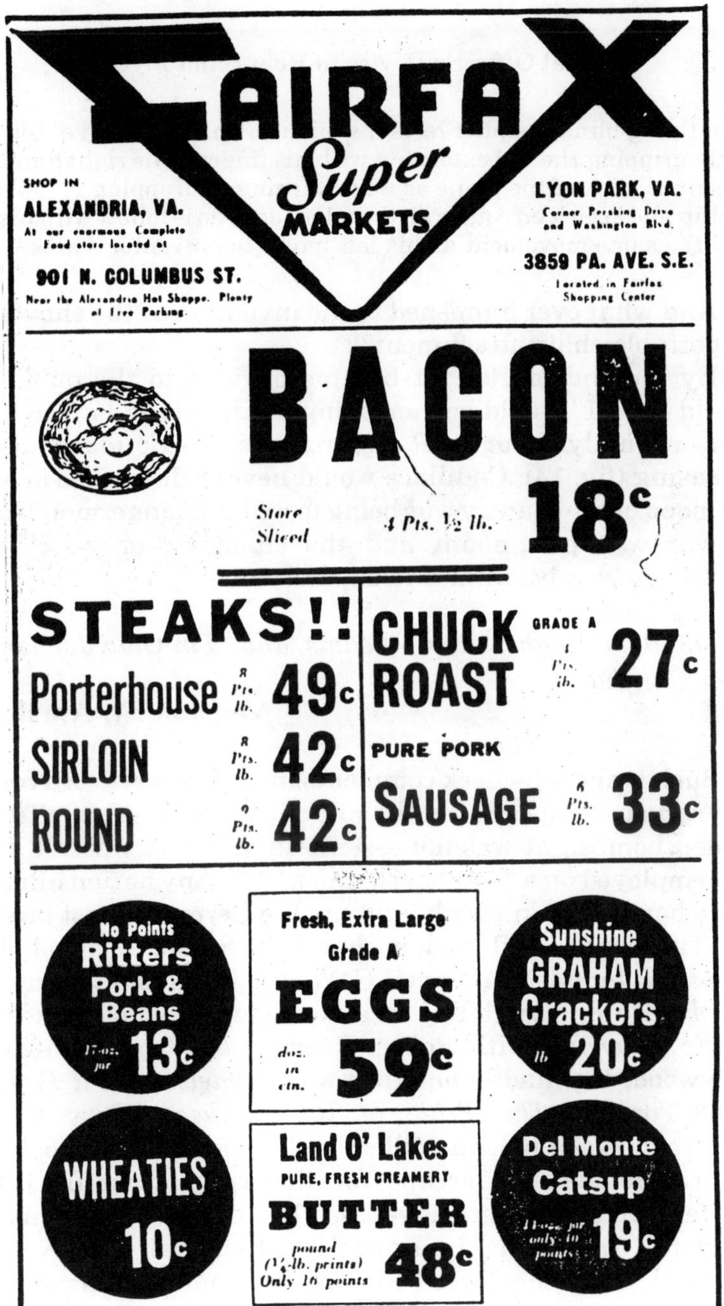

Fig. 13

Washington *Post*, weekly advertisement.

"Naval Officer's Device to Help Blind Flying"

When flying blind the pilot rests his left arm on the second or top plate, lightly gripping the rod extension with his fingers. His right hand grips the control stick of the plane as when turning or dropping .... To right the ship and keep it on course all the pilot has to do is follow with his right hand the same movement as his left hand, the inventor states.[17]

And what ever happened to the invention of the shovel with the portable chair attachment?[18]

By the end of 1941, it became obvious to the public that "World War II" would be increasingly difficult and much longer than originally thought. Rationing of strategic materials was increasing (fig. 14). Cadillacs would never sell for $1,345 again. The mood of the country was being forced to change, mobilization for war was paramount and the emphasis on superior air technology was beginning to expand.

ii. 1942: Bomb Sights, Radio Beams, and "I'm Glad Joe Louis Is On Our Side"

***Newsweek,* April 1942**

Such tongue-in-cheek comments in the press reflect a concern for a serious perspective in the role of technologies. In effect, the Norden bombsight was not essentially a military secret. It had been employed on a "war basis" since 1935. Any nation interested in the bombsight knew about it, as the Germans most probably did. The movies reflected a change in national mood: *Eagle Squadron, Invisible Agent* (H.G. Wells), *Wake Island, Spy Ship, This Is The Enemy* (billed as a Neanderthal who smoked) and *Pacific Blackout* (with Robert Preston) all demonstrated that Hollywood, too, had gone off to war. *Somewhere I'll Find You* (Gable-Turner), *The Pride of the Yankees* (Gary Cooper), *Orchestra Wives* (Glenn Miller) and *Holiday Inn* (Crosby and Astaire) provided little escape from the reality of war (fig. 15). The new fashion styles were advertised by Lord and Taylor as "The Beautiful Invasion: Imaginative, Pictorial. So New! So Beautiful."[20] In the Tweed Shoppe one could purchase a four-season hand-loomed tweed coat called the "P-40 Bomber."[21]

And the technological developments also reflected an all pervasive wartime posture. UHF radio frequencies were being used for instrument approaches. A suit for dive bomber pilots of

Medals for Misers: ***This advertisement, originally prepared by one New York City store, appealed to eleven other large metropolitan merchants swamped by hoarding shoppers. So they joined in sponsoring the ad in newspapers last week. As The New York Times put it: "If it is news when a man bites a dog, it is certainly news when a merchant urges a cus-***

Fig. 14

New York *Tribune*, 1941. Also appeared in *Newsweek* for national circulation.

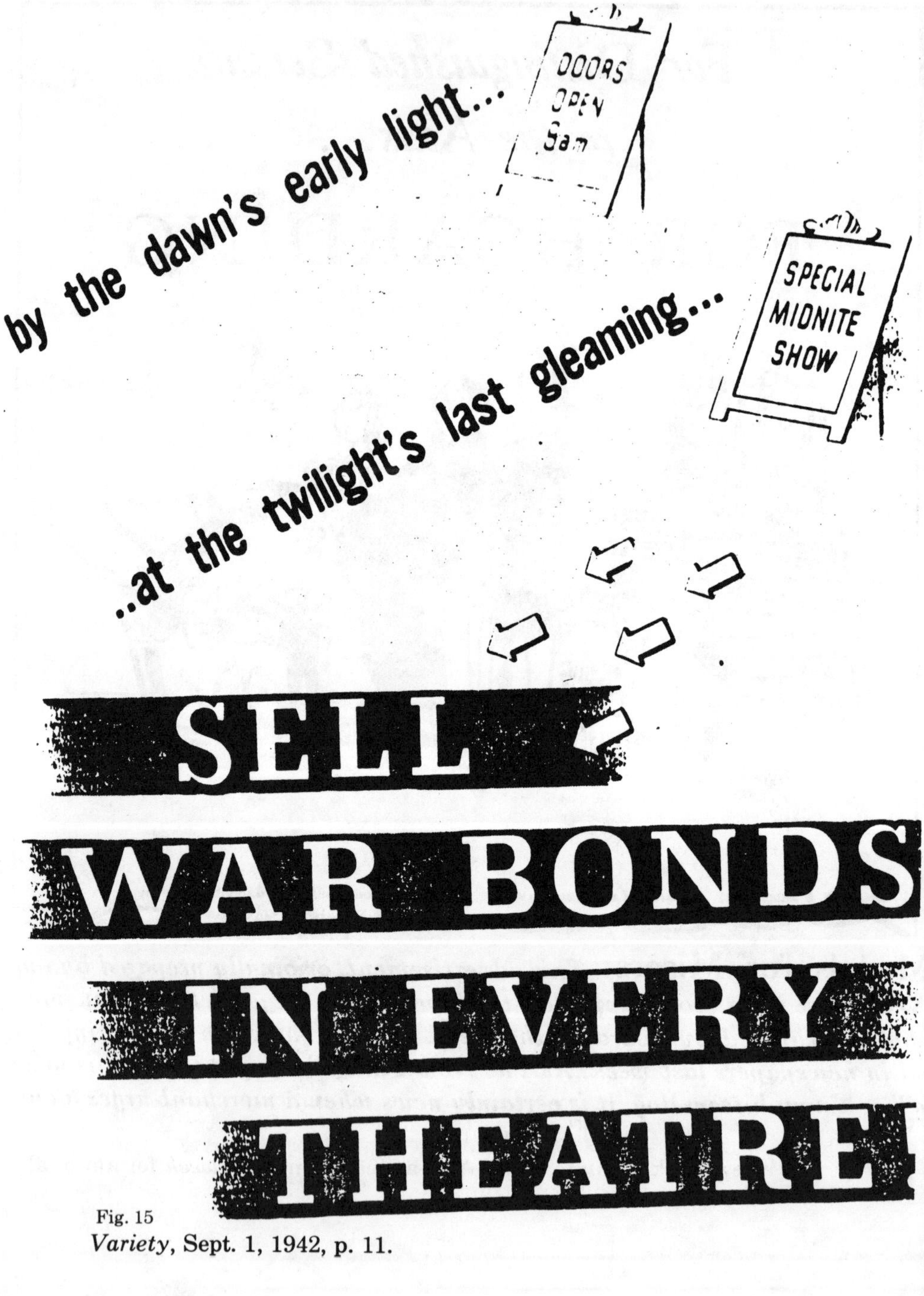

Fig. 15
Variety, Sept. 1, 1942, p. 11.

inner elastic and an outer inelastic shell was being used. Non-explosive gasoline tanks, built-in parachutes, wind drift bombsights, hyperbolic landing systems, centrifugal oil separators, radio locators for spotting planes, wireless heated pilot units, flight recorders, fuel efficient propellors (plus 5%), ground to air altimeters and retractable pontoons all attest to the accelerated pace of technological development reported in American newspapers.

Of course, censorship and secrecy demanded that details not be made public. In fact, researchers were so notified:

> Research scientists will henceforth be unable to obtain scientific equipment without the approval of the Director of Industry Operations.[22]

Also, patents assigned to foreign nations were placed in the offices of the Alien Property Custodian as was the case with one bombsight developed by Italian A. Baroni (Rome) whose device made corrections for ground speed, air speed, and slip—so, too, with an air cooled line engine. One Romeo A. Bossi patented yet another bombsight kept secret because of existing military use, but "that the present invention discloses nothing already generally known."[23]

The Victor Adding Machine Company increased from six Norden bombsights a day to three hundred a month, and the sight was improved to permit precision bombing from greater altitudes.

By July and August of 1942, the press was reporting that some of the technologies had peace-time applications (fig. 16). Whether such comments were made to justify experimentation and expenditure is not altogether clear, but reports, such as that of a revolutionary plane finish, are indicative of many similar stories:

> It is especially designed to add a minimum of weight to the planes and gliders. A wide use for the finish is seen in peace time, although it was especially developed for military purposes.[24]

Plans were also being made for surplus airplanes as early as July of 1942, especially in reference to triplicate radio beams designed to provide invisible "tracks" used as directional aids.[25] Such optimistic reports challenged the Japanese view of American technology which claimed "the war was just beginning," especially said in reference to a 10,000 mile long

Fig. 16

Surprisingly, a great many advertisements of this type projected a post-World War II lifestyle.

range superfortress which the Japanese claimed the Americans had just developed.

Perhaps cautious optimism best describes both the technological developments of 1942 and, in general, the tone of the United States as reflected in the press by such reports as the following:

"Louis Signs at $21, Eager to Get Japs"

The toughest fighter in the world became a soldier of the United States here yesterday as Joe Louis, whose "battling" fists have brought him $2,263,784 and the world's heavyweight championship, enlisted to fight for his country at $21 a month.

"Them Japs is all lightweights, anyway. They don't have any heavyweights."[26]

Jack Dempsy also volunteered, but was rejected because of his advanced age. Gene Tunney gave the soldiers some advice, which must have left some GIs smiling: "Jumping rope is better than smoking."[27] When hearing of Joe Louis' enlistment, one raw recruit (unidentified, of course) was supposed to have remarked, "I feel better now with that guy fighting for us."[28]

iii. 1943 "Lean but Clean," Air Supremacy, and Mechanical Gibson Girls

The tentativeness of 1942 vanished almost instantly as 1943 began. The press reflected a typically optimistic American attitude of recognizing and accomplishing stated goals (fig. 17). This outlook was fostered to a great extent by war news which indicated the fighting on all fronts was turning toward the Allies' advantage. "MacArthur Fliers Destroy 22 Japanese Ships," "Enemy Losses [sic] 15,000 Men in Convoy, 55 Planes" and other war reports were typical of the daily headlines.[29]

As increasing belief in the superiority of Allied technology was proclaimed repeatedly: U.S. bombers were declared superior; mechanical devices on planes provided the pilots with freedom for important functions;[30] the Flying Fortress was said to have destroyed one hundred and fifty-nine Zeros before the Japanese discovered it carried tail guns;[31] and, phosphorous was touted as being helpful in allowing pilots to read instruments in the dark. The list goes on and on, but the folks back home were more than just informed of events and technologies, they were *included* in the war effort (figs. 18-23).

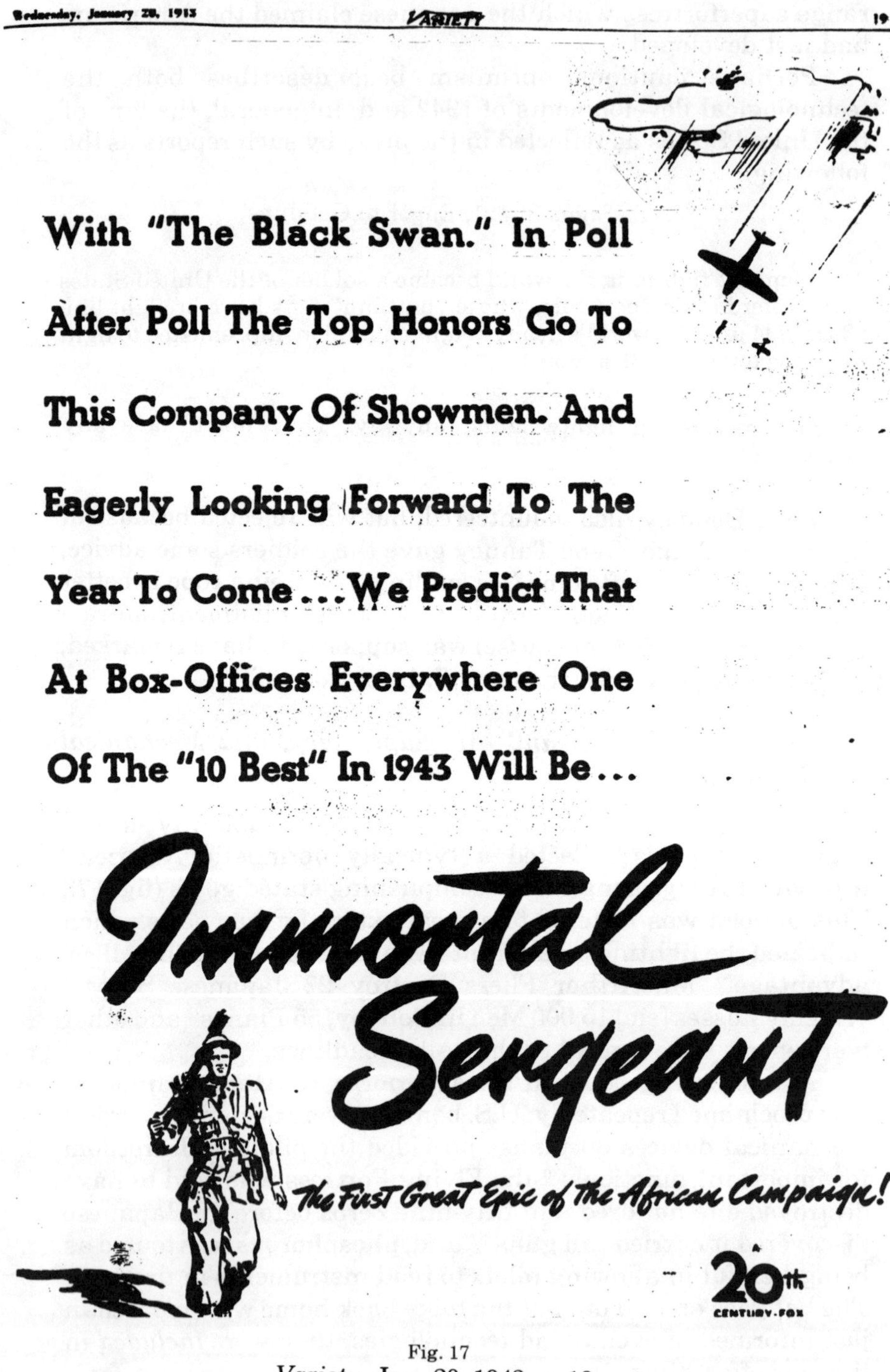

Wednesday, January 20, 1943 VARIETY 19

With "The Black Swan." In Poll After Poll The Top Honors Go To This Company Of Showmen. And Eagerly Looking Forward To The Year To Come ... We Predict That At Box-Offices Everywhere One Of The "10 Best" In 1943 Will Be ...

Immortal Sergeant

The First Great Epic of the African Campaign!

20th Century-Fox

Fig. 17
Variety, Jan. 20, 1943, p. 19.

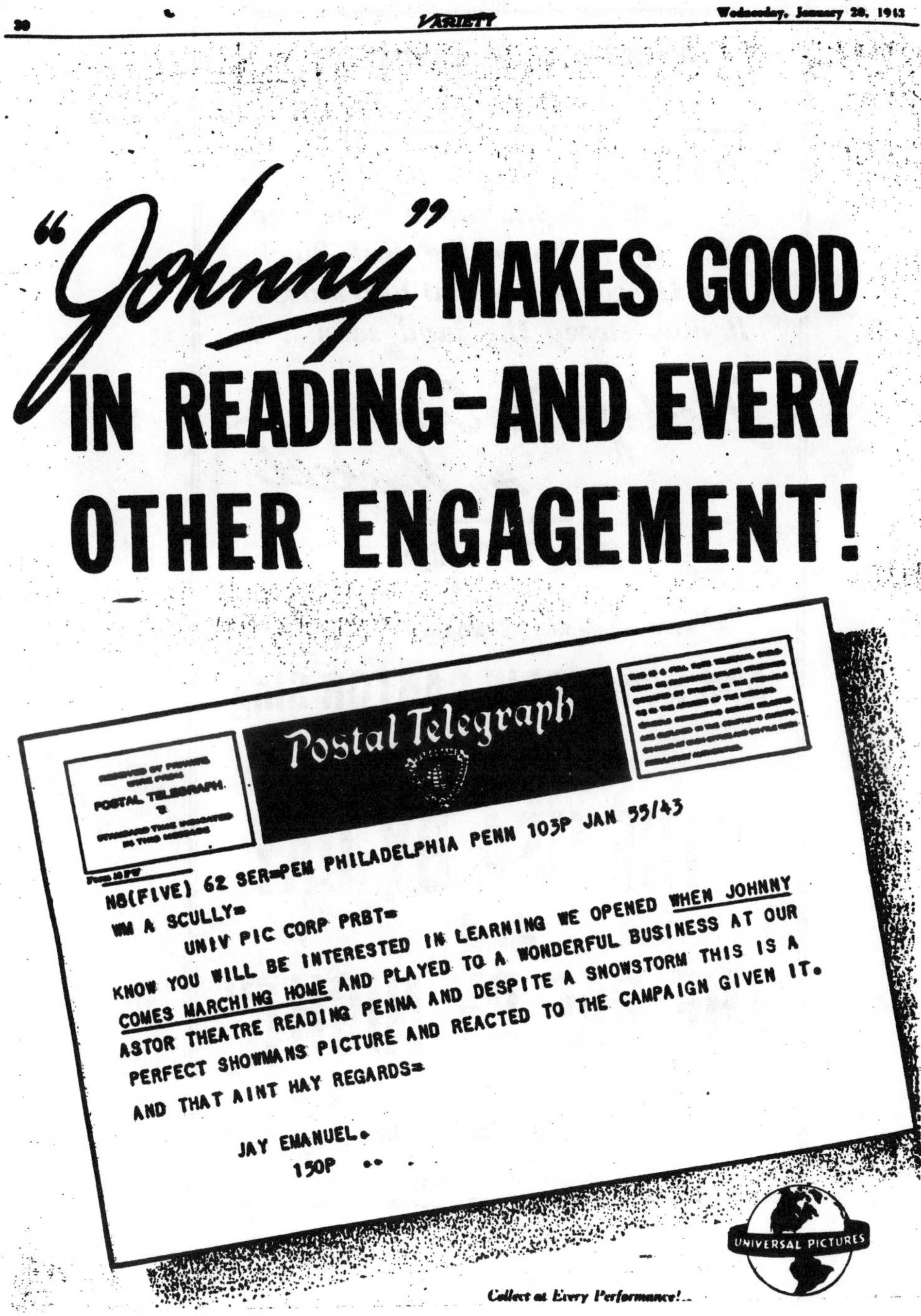

Fig. 18

Variety, Jan. 20, 1943, p. 30.

Walter Winchell
In New York
"The USA by Day and the RAF by Night" is the name of a new song by radio scribe Hal Block and UP correspondent Bob Musel. It may sweep the land soon. . .
Thanks for the Orchid
Mr. Winchell
Tune In To-Night
June 14
WEAF and the RED NETWORK, 9 P. M., EWT
And Hear EDDIE CANTOR Sing
A Great New Song
"THE USA BY DAY
And
THE RAF BY NIGHT"
PARAMOUNT MUSIC CORP.
1619 Broadway, New York
CHARLIE ROSS, Prof. M'g'r

Fig. 19

Fig. 20

Variety, May 12, 1943

Fig. 21
Variety, March 24, 1943, p. 39

Fig. 22
Variety, July 7, 1943, p. 31.

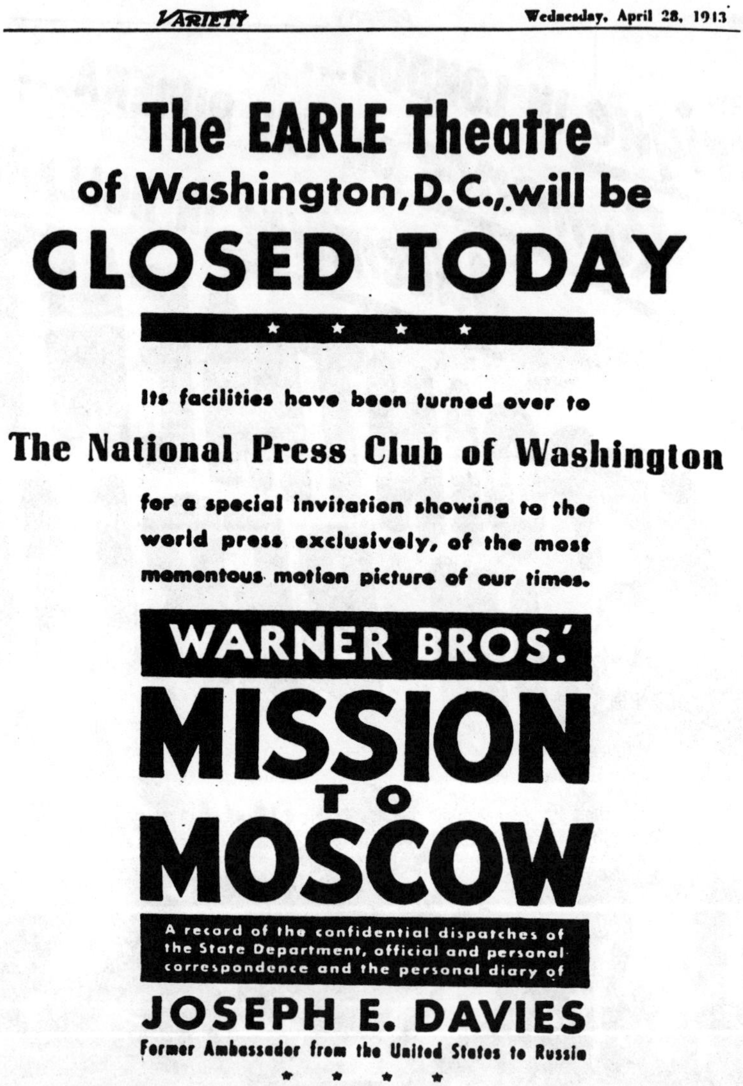
VARIETY Wednesday, April 28, 1913

The EARLE Theatre
of Washington, D.C., will be
CLOSED TODAY

Its facilities have been turned over to
The National Press Club of Washington
for a special invitation showing to the world press exclusively, of the most momentous motion picture of our times.

WARNER BROS.'
MISSION TO MOSCOW

A record of the confidential dispatches of the State Department, official and personal correspondence and the personal diary of
JOSEPH E. DAVIES
Former Ambassador from the United States to Russia

Fig. 23
Variety, April 28, 1943, p. 26

Wednesday, April 28, 1943 VARIETY 37

from the

124 stations of the Columbia Broadcasting System ★ ★ ★

Fig. 24
Variety, April 28, 1943, p. 37.

"Industry Has Part in Air Supremacy" read one headline in reference to the development of one of the most effective secret weapons of the war—100 octane super fuel, and "no Axis planes known to our military authorities are equipped to use it."[33] Truman hired forty-two women from Columbia University to be engineering assistants in the total effort, as the press quickly informed the public.[34] The Flying Fortress was praised as a "Seattle Child": "Everytime one of them smashes an Axis installation, it gives new meaning to the banners at the Boeing plant 'Builders of Victory'."[35] In another plant, three hundred beverage container workers were switched to making high altitude oxygen bottles "with no loss time to the men and no idleness of machines."[36] Civilians were advised to lead "lean but clean" lives in order to double aircraft production, since the "Arms Goal in 1943 [sic] Twice 1942 Output; Stress on Planes."[37] Patriotic blonds were growing fourteen-inch tresses for precision plane instruments; "Girls Grow Curls To Help Win War"[38] (fig. 24). "Surgeon is Honored for Flak-Proof Vest," a vest designed by one Col. Malcolm G. Graw which saved lives and "increased morale."[39] The glorification of the middle-class man was not lost in the following article about dictating machines placed on planes: "Like a busy office executive, United States Air Force men can now dictate their observations."[40]

There were, however, significant technological developments which indeed did contribute to the war effort, and in some cases, even to post-war non-military aviation. The Gibson Girl (an hourglass shaped radio transmitter) sent out rescue signals from downed flyers which covered approximately 100,000 miles; in fact, the Fraternal Order of Gibson Girls was formed by those who were saved by employing the device.[41] Low carbon steel was being used in training planes. Wind tunnels had been developed which could produce 400 mph winds at 67° below zero, thereby passing 4,000,000 cubic feet of air over the test subject every minute.[42] Sealed machine guns allowed firing at high altitudes, and rocket boosters signaled the coming of jet-propulsion. Improved plane tires allowed for high speed landings, the public was told.

A British bomber, the "Mosquito," was said to owe its high speed and strength to "a smooth, light flexible wood skin," which was claimed to be stronger than metal.[43] Readers across the nation were informed of the existence of special anti-submarine armament placed on Navy bombers; however, the details were to remain secret. The U.S. Army developed an anti-aircraft gun which could hit planes flying at 60,000 feet, said to be one of "the most powerful weapons forged from American ingenuity."[44]

Perhaps the technological development was best summarized by General G.M. Barnes, who said:

> We now have the best weapons in the world, but we are not satisfied, and if the war lasts two years, the weapons used then will only slightly resemble those used at present.[45]

However, General Barnes' appraisal was not accepted without challenge. The New York *Times*, in editorial response, stated: "We can take pride in American ingenuity as well as in American courage .... But the further we go, the surer we had better be that there are not more wars."[46]

Such admonitions were invariably lost in the mish-mash of cheerleading and advertising which accompanied the optimism of 1943. The American poet Carl Sandberg encouraged people to buy more war bonds since "Wishes won't wash dishes."[47] The fashionable colors in January of 1943 were red, navy blue, brown and gray. Ethel Merman's "smash hit" of 1943 was entitled "Marching Thru Berlin." Citizens were asked to give their shirts or lose them (fig. 25). One former frying pan company advertised that it was now making airplane parts, and housewives were asked to save waste kitchen fats which were needed in the production of glycerine. The WAVES were increased fifty percent to 25,000. American readers saw the following from the National Coffee Department of Brazil:

> Coffee is the greatest stimulating beverage for the soldiers fighting to preserve democracy and liberty.[48]

The movies, too, became more optimistic as the tide of war changed: *Assignment in Britany, Casablanca, Air Force, Edge of Darkness, Desert Victory, Victory Fleet, Keeper of the Flame, Immortal Sergeant* and *Hangmen Also Die* (Brian Donlevy) all offered hope for victory (fig. 25-28). It must have been somewhat refreshing to know one could wear a white hat, white gloves and a navy blue dress or skirt and dance to the music of Ozzie Nelson and his orchestra—in person!

The technology of 1943 consisted in part of a hodgepodge of non-rubber adhesives, fireproof clothing, "How to Serve Meat When You Can Get It," bird-proof windshields, electronic de-icers, flying control synchronizers, rocket firing guns, plentiful flowers

Wednesday, November 8, 1944 VARIETY

Make Your Public Remember Pearl Harbor With

Free Movie Day

DECEMBER 7th!

WITH PATRIOTIC FEVER running high, everyone will be eager to "Remember Pearl Harbor" with a bond. And Free Movie Day is the event to give your campaign a running start for the last week of the Sixth War Loan Drive.

As you remember, Free Movie Day was tried on a nation-wide basis for the first time in the Fifth War Loan Drive. More than eight thousand theatres participated in a simultaneous bond-selling event that turned out to be the sensation of the drive. It's been tested and proven tops, and now it's tied to one of the biggest, one of the most important, war dates in our history. The results can surpass anything the Motion Picture Industry has ever done for the Treasury Department if you will lay plans now to capitalize on the nation-wide publicity that will urge every American to Remember Pearl Harbor with an extra War Bond.

2-AB

Fig. 24a
Variety, Nov. 8, 1944, p. 13.

Fig. 25
Variety, April 28, 1943, p. 24.

Fig. 26
Variety, May 26, 1944, p. 12.

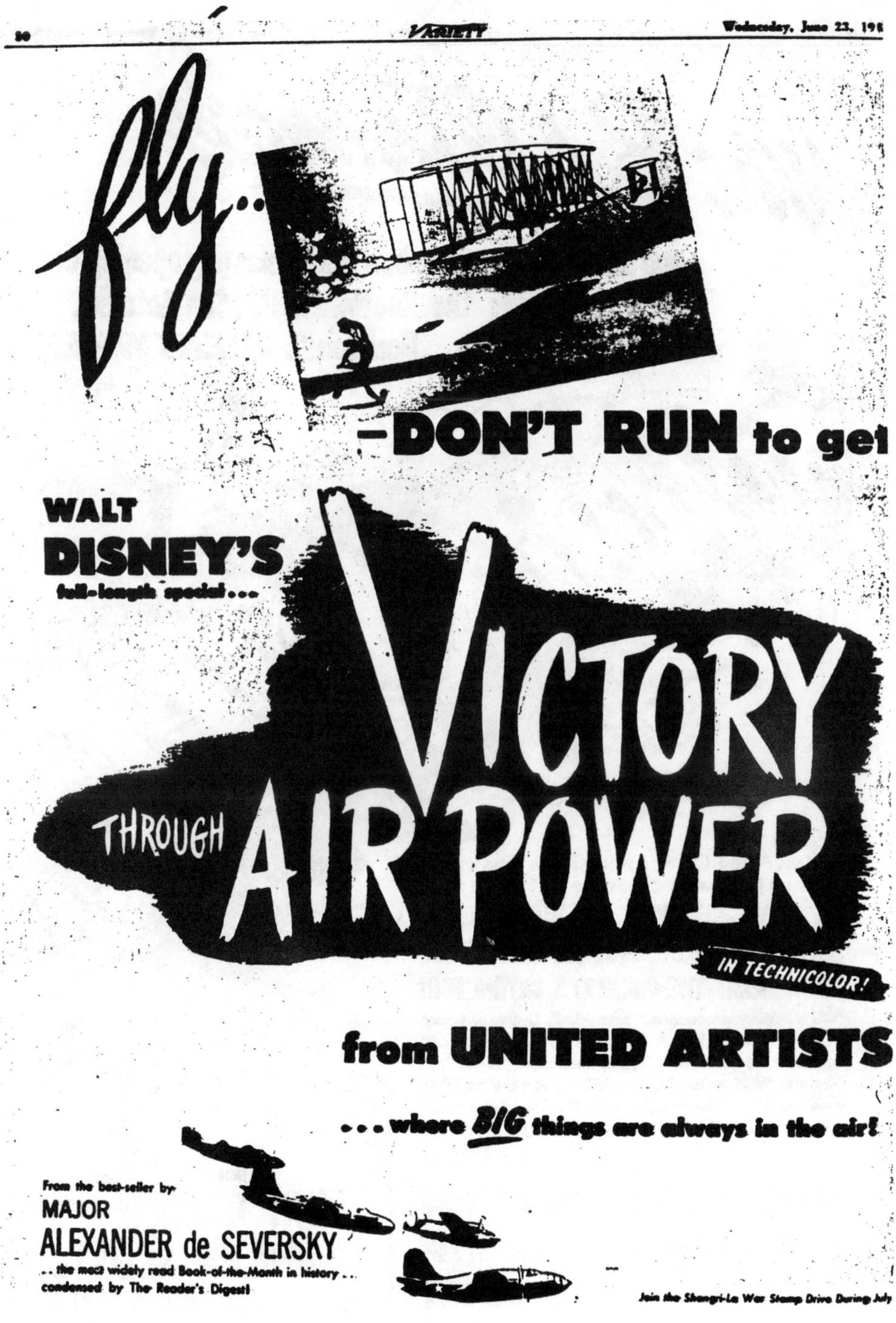

Fig. 27
Variety, June 23, 1943, p. 30.

Fig. 28
Variety, May 26, 1943, p. 24.

at lowered prices, and "A Pound of Sausage Meat Serves Six" (Fig. 29). The *Science News Letter* suggested that alkalized beef plasma may be safe for transfusions, explained that a heated ironing board was being developed, and questioned if breaking the four-minute barrier for the mile was realistic.[49]

Such suggestions for technological improvement were so blatantly inane that the humor was not lost on the press.

A South Pacific Base, February 6 (AP). Dump a general in the drink, storm toss him for six nights and five days, starve him and up he comes smiling with sound recommendations for the safety and comfort of others who are similarly marooned.[50]

His suggestion? Watertight watches so that those aboard will know what the hour is!

The press indicated what all technocrats come to know sooner or later—that no technology is forever and that technological secrets are at best very seldom kept for very long: "The Norden bombsight went the way of practically all wartime secrets last week, bearing out the rule that the enemy catches on sooner or later."[51] Such statements not only informed the public of technological events, they also implicitly suggested that only with more technology, with more technological advances, and with even better security could the war be more quickly ended.

However, the technological euphoria which had developed during the first nine months of 1943 soon became tempered by the following news on September 22, 1943:

"Churchill Reveals Nazis Using
Glider Bomb Powered by Rocket"

To American and British Navies the rocket bomb is dangerous but not mystery. When its mechanism is completely understood, it will only be a matter of time before the Allies conquer it.

Two days later, this story appeared in the same newspaper:

So far as remote radio control is concerned, what patent lawyers call the "art" was developed long ago .... There can be little doubt that the bomb has been provided with wings and rudders—a provision which hardly demands exceptional ingenuity.[52]

Was the American press "whistling in the dark"? It seems so. Indeed, in retrospect the glider bomb was primitive when

Fig. 29

Food advertisement with rationary coupons constantly keep the war shortages in one's consciousness.

compared to the V-rockets which followed in 1944, but at that stage of the war and at the then known level of pilotless aircraft technology, the simple glider bomb was the first of many technological developments which "protracted the night."[54] American readers were informed that a "Radio Controlled Bomb is Unlikely to Help Germans"[55] and that concern is unwarranted since "as Mr. Churchill pointed out, 'we have always been able to find the answer to any new problem presented'."[56] The lack of specific official reporting of the glider bomb intensified the concern and added to the confusion. Such information as the following did little to encourage or enlighten:

> This "flying bomb" seems to be launched from a plane—It apparently has some sort of rocket propulsion—It is either radio controlled or some sort of magnetic device.[57]

Within eight days of Churchill's original announcement, what seems to be the truth of the situation was noted: "The development of rocket propulsion, aerodynamics and electronics has made almost anything possible."[58] The *Science News Letter*, however, took a less objective view of the German technological development by claiming that cross-channel rockets were so rumored that any discussion would be speculative and premature.[59] The "robombs" were not rumors to one U.S. Army sergeant who tells there is "no place to run to," so one must stand starry-eyed, with tremors, with whiteness of face, and with beads of perspiration on one's forehead waiting for the "robombs" to come,[60] yet Sergeant Adolf Woltman's experience was but a mild prelude to the arrival of the devastating V-rockets of 1944.

iv. "And just what the bloody hell else do you think anyone can do? You'd think we had some choice in the matter!"[61]

These words echo the sentiments of at least one British resident who experienced the prolonged and intensive V-rocket attacks which began in June of 1944. Such resignation, however, did not characterize press reports earlier in that same year.

On January 3, 1944, *Time* reported that cross channel bombing and U.S. precision bombing on Norway had greatly reduced Nazi "rocket gun effectiveness."[62] *Life* relayed much the same by telling how three thousand Allied planes had bombed the French rocket coast and had destroyed rocket gun installations:[63]

"The rockets will never become 'super weapons' which some Nazi propaganda had tried to make of them."[64]

Ironically such reports were proven accurate only because the Germans had made a quantum technological leap to the V-rockets by June of 1944. Readers of the *Ann Arbor News* (Michigan) found that "Britons Prefer Bombing to New Mystery Plane."[65] They read how one eighty-year-old English woman commented that the "mystery plane" seemed "horrible queer to me" and how a delivery boy claimed, "I don't know why but it's sort of ghostly."[66]

If technological success can be measured by the amount of response generated, the V-rockets were overwhelmingly successful. One New York *Times* editorial made an effort to diminish the German achievement by claiming that "apparently we have nothing more than an aerial torpedo" which the Germans use "because of their [sic] advertising value."[67] The Allies, readers were told, had experimented as far back as 1935, but had decided not to use such rockets because of a preference for hitting specific targets.[68] In part, the editorial half-praised the achievement: "We have no desire to belittle the German effort. It indicates a willingness to depart from conventions not usually displayed by general staffs."[69] The Chicago *Tribune* was less indirect when it claimed that "weird craft" was really designed fourteen years earlier by E.E. McDonald of Zenith Radio Corporation.[70] One columnist attempted to ridicule the "robots": "Hitler has just thrown his fanciest punch at us . . . the noise was like a lawn mower working at too high a speed."[71] In a column the following day, he wrote of one attack: "This was Derby Day and everyone in England would have laid odds that Germans would try to spoil it. They did."[72]

Readers around the country read how the "infernal machines" flew, that one should "duck" as soon as possible when one heard the "whizzing," and that the Germans had no real military objective in mind but were simply exercising revenge upon England. If anyone knew anything concrete and accurate about the V-rockets, they were not telling the press. In fact, the only defense against the "wonder weapon" seemed to be a wall of propaganda: "Whatever the Germans may have, it seems doubtful it is not anything previously experimented with by the United Nations."[73] The London *Times* took a rather stoic viewpoint of the attacks by claiming there was "no reason to think that raids by this weapon will be worse than, or indeed as heavy as "those already experienced."[74] Indeed, the *Times* clarified the raiding as "nuisance raiding" instigated by "the calculated brutality of the German militarists" (fig. 30). Help was on the way, claimed the Washington *Post*, because Allied scientists "for a long time anticipated that Hitler would try a trick of this kind."[76] Therefore, Allied scientists have been "busily at work discovering ways and means of counteracting it."[77]

"If this is the best Nazi genius can devise, their miltary power will go out like a Roman candle," said an editorial in the Detroit *Free Press*.[78] "Hitler's new device resembles an enlarged steel model of a child's Fourth of July rocket, equipped with wings and explosives," said one front page story in the New York *Times*.[79]

Propaganda on both sides increased as the number of V-attacks continued. The American public was told "Britain Perfecting Antidote As Robot Plague Increases,"[80] "Coastal Guns Exploded Nazi Robots in Air" (from which they "Fall Like Clay Pigeons"),[81] and "Nazi Propaganda Soars With Rocket."[82] The V-rockets were called the logical weapons for the Germans to use since they did not control the airspace over the European theatre, since they could not afford any close combat and since they were prone to revenge the June 6 D-Day invasion.

"Nazis Losing Again in 'Nerve War' " since the robots do little more than arouse curiosity, some readers were informed.[83] "German People Deluded Fantasies of the Flying Bomb,"[84] and "The Secret Weapons no surprise to readers of this magazine,"[85] were typical of many reports which appeared in the press.

But the propaganda could only counter the technology to a point and the censorship could only conceal a small part of what was happening during the V-bomb attacks, especially with the advent of the V-2. The Chicago *Tribune* reported that "Nazis Use New Type Bomb to Hit Britain,"[86] German war reports indicated that London had been bombarded by V-rockets for one hundred and sixty hours without interruption.[87] Was British propaganda designed to soften the attack by shaving the hours to one hundred, or less? How could the propagandist counter the reports that Spitfires faded like gliders in the pursuit of the large jet propelled missiles which traveled at over 600 miles an hour?[88] Even the names assigned to the weapon by each side soon lost whatever effect they had solicited: The German names of "Hell Hound," "Hell Dog" and "Dynamite Meteor" were countered by the English terms "Winged Comet," "Buzzbomb," "Bumble Bomb," "Flying Bomb" and "Whirly."

By June 30, 1944 the semantic battles seemed exhausted, and

Fig. 30
Variety, Jan. 13, 1943, p. 11.

the Allies realized that a threat did indeed exist from the new technology. Perhaps General Fuller, the military commentator for *Newsweek* presented the most objective appraisal: "He [the enemy] knew that we had landing craft, yet he could not meet them, therefore he was surprised. Now we have been surprised." (He also suggested that the official reaction was to minimize the effect of the pilotless plane.)[89] The V-rocket reports became much like other war reports: "Bombers of All Types Rip Nazi Attack."

The V-rockets were real and the effects were obvious: "two in sports crowd killed"; a hospital was hit, another landed in a convent (killing one nun).[90] One American scientist called the rocket bombs "a development as revolutionary as the airplane, with a tremendous future and a definite political problem."[91] By June 30, 1944, the effective reality of the V-weapons was reinforced when the British asked for uncensored data on "robot deaths" since newspapers had not been allowed to publish vital statistics or photographs of the damage inflicted. Yet most statistics were not released until August of 1944. In seven weeks of V-bombs 4,735 persons were killed by 5,340 of the weapons, 14,000 were wounded (nearly double the Blitz) and 817 homes were razed or damaged (as compared to 2,750,000 for the first four years of the war).[92] Over 2,000,000 people had been uprooted.[93] 83,000 workers were employed in repairing the damage, 150,000,000 tiles and slates were needed, as was 200,000,000 square feet of ceiling and 50,000,000 square feet of glass—according to the Ministry of Reconstruction.[94] The V-rockets were indeed formidable war technology, as the press finally reported.

However, no statistical analysis could adequately portray the human toll in terms of fear and death. In the first place, there seemed to be something basically dishonorable about the use of the "blasted things." Michael Young wrote: "In restrospect, planes with pilots seemed relatively civilized. You could shoot at the pilots, who were men of flesh and blood. Pilots at least tried to aim at their targets."[95] Sedatives were used for tension, beer for cheer. Restaurants were often empty. Radios were not used because the programming would cover the sound of incoming missiles. Crowds thinned, except it was claimed—for soldiers and girls together in packs. "And so it went on. And the bloody things kept snorting over and smashing down over the lot—some far some near."[96] One staff sergeant, United States Army Air Force, wrote the following in a letter in the *New Republic*: "They are terrible. I have some idea of explosive power, and I have never been so angrily frightened."[97] The British Minister of Home Security is quoted as having said: "We may have to put up with quite a bit of it yet (as of July 24, 1944), despite everything that our airmen and our armies can do to defeat it."[98]

Headlines began to appear which gave testimony to the magnitude of the weapon: it is thirty feet long, flies at sixty miles altitude in the stratosphere, and it descends at 900 miles an hour; "Apples were shaken from trees when a V-2 landed in an orchard;" and "rescue workers stand halfway down enormous craters gorged by V-2."[99] "You just cower under the bedclothes," wrote Phyliss Bently for the *Saturday Review,* "or behind furniture to keep away from glass...you feel ashamed because somebody else has died and you haven't."[100] Some Londoners simply resigned themselves that the bombs would indeed come and that there was very little that they, as individuals or in concert, could do about it. Some simply followed this advice: "If doddle dollies don't dawdle. DIVE!"[101]

> Londoners shaken but still full of understatement, talked mostly about the lucky misses. One bomb sailed close over a crowd at a band concert in Hyde Park, landed a scant 500 yards away, killed only two. The band went on playing *On Steps of Glory.*[102]

Such was matched only by the rational irrationality of the following determinist:

> It must have taken the Germans a lot of trouble to make the bloody things, and then they have to get them into those pits and up in the air and it is quite a long way from France to London, and if they do get to London, they have to find Hackney. And even then, it isn't everyone who can find Bulstrode Road, and if they do, ten to one I am in the pub.[103]

Still, such bravado is probably more indicative of concern than of courage, and the British government did little to officially inform the populace for fear that Germany would be aided by the same information (fig. 31). Since there was no official word, there was no indication of how long the attacks would last, or whether they would diminish or intensify: "Nothing came from official quarters to dispel Briton's concern that they would hear it (roaring and thundering) awesomely multiplied if the war did not get over in a hurry."[104]

v. "Boomerang" or the "Weapon of the Future"

Besides civilian reactions to the V-attacks, two other

Fig. 31
Typical of Period Posters

identifiable positions were often stated in the press: one, there would be no retribution for Germany's attempted revenge; and, two, there was the technocratic view that V-weapons would change the course of warfare.

Those calling for revenge on Germany, political and otherwise, found the V-weapons provide an adequate base for reaction. The V-2, for instance, became adequate testimony to "Nazi meanness, cleverness and persistence," and the attacks were as "lurid as something out of a Wells novel."[105] One writer claimed that "Americans had better know that on this side of the Atlantic a once kindly distinction made by us [England] between Nazis and Germans has gone."[106] "The Flying bomb has accomplished one thing," said another reporter, "It has revived the feeling of hatred which died down after the blitz."[107] Calls for action ranged from being "kind first and last"[108] to "justice without mercy ... the only hope for a better world."[109] After a visit to a buzz-bomb sight (where 5,000 Russian, Belgium, Dutch, French and Polish "slaves" worked) one commentator said: "I left this devil's workshop, as I left many a Savoy poker game, feeling that our enemy in this war is at least half fiend."[110] Every thinking Englishman had been so shocked that it was virtually assured that "Germany will not enjoy an independent existence during the foreseeable future."[111] Germany could no longer be trusted to conduct war in an honorable manner, certainly not after one V-attack where "nine children at a birthday party were blown to bits."[112] "Considering Germany's industrial plights, V-2 might seem a forlorn hope...but after the V-blitz, Britons were not discounting anything,"[113] read one *Time* article, especially since most Londoners "still felt unsafe from the last fling of German vengeance."[14]

The technocratic view reflecting the V-weapons as the weapons for future warfare concentrated far more on the technology and somewhat less on the political vengeance than did the revenge seekers. Most scientists assumed the same position as did some militarists—that the robot planes were "not new." In fact, *Science News Letter* reported that a patent for an aerial torpedo was applied for on April 25, 1919 and granted on April 5, 1927 to Dr. Charles F. Kettering (Patent No. 1,623,121).[115] The report goes on to state "It seems that the only new thing that the Germans have done ... [is] jet propulsion, which by the way was first invented by an Englishman."[116] An earlier report in the same periodical reported that the secret of the V-2 auto-steering system may be in U.S. Patent 2,351,977 (with the Alien Property

Custodian) filed on April 1941 by Adam Kronenberger and Fritz E. Bartelt of Berlin, Germany.[117] Because of censorship, information available to the technocrats was limited; however, Ernest Hemingway gives an interesting perspective on how some reporting was done:

> Writing under censorship is necessary and proper in time of war, and we all censor out ourselves everything we think might be of any possible interest to the enemy. But in writing about the air on the basis of trying to include color, detail and emotion, there is a certain [sic] analog [sic] to sports writing.[118]

"Battle of England: The Score for the Robots" read one *Time* headline, which, besides giving damage and casualty statistics (basically the same as those already mentioned), informed the reader that only 12 1/8 percent of the 2,800 V-weapons were destroyed by Americans.[119] One boxscore was reported in *Newsweek* as being of a "typical day": of 143 robots to coast, 35 to RAF, 17 to barrage balloons, 65 to electric gun direction, and "only 26 got through,"[120]—that the electric gun director was a U.S. invention was proudly proclaimed in the same issue. The *Time* story also contained one "long guarded secret—that 92 percent of all casualties were in London":[121]

> If another war starts, Britain and Germany may make shambles of each other's cities any morning at the push of a button. For that matter, if the robot is improved as much as the tank and plane were between wars, one end of the shambles might be New York.[122]

The "Buzz Bombs" were heralded by some as being so advanced that if they had been used two weeks earlier the D-day invasion "might well have failed."[123] "The Germans could boast that they had raised the art of killing to the highest peak of science and efficiency";[124] however, the fatalistic resignation of populations which results from the lack of warning by the V-2s classifies the V-2 as a propaganda weapon as well as a military one—even if it did lack the suspense of the V-1.

The same claim of civilian resignation was later made about the atomic bomb. It seems both the V-weapons and nuclear weapons do indeed represent a quantum leap in military technology. It appears the consensus that both types of weapons had been developed too late for effective World War II use.[125] In fact, *Life* suggested that the robot bomb, although ineffective "now," may be "the way the next war will begin."[126] Both the Russians and the Americans viewed the "trend of the robot" as the subject of what some super-robot might eventually do for them.[127]

The V-weapon technology and the future is perhaps best summarized by Major General J.F.C. Fuller (C.B., C.B.E., D.S.O., British Army Retired) who claimed: "It is self-evident that with such a revolution comes the doom of the bomber, of long range artillery and probably the cannon in most of its forms."[128] All of which lead him to the assumption that "a war of maximum annihilation" could occur "without moving a man."[129] Rocket bombs could not only change the face of this world but of all worlds since in a technological sense Fuller claims, "we live in extraordinary times, in days of strange and violent possibilities."[130] J.B.S. Haldan was much more optimistic in his appraisal of V-rocket potential: "For the cost of a day of war, it should be practical to send a series of rockets round the moon and photograph its far side."[131]

vi. Lights On!

The lights were not the only things turned off by the V-attacks; it seemed that human judgment also dimmed at times. *Science News Letter* reported that one specific type of plant folded its leaves during the V-2 blasts:

> It must be added that this plant that can't 'take it' is not a native English species, but an immigrant from the Indonesia tropics (commonly known as the telegraph plant, Desdondium gurane).[132]

This unpatriotic foreigner also assumed "a sleeping posture whenever an explosion goes off."[133]

"Now it can be revealed how one decisive RAF raid held up production of German robot bomb," claimed an article in *Reader's Digest.* The story, originally in *Toronto Saturday Night* (Sept. 2, 1944), relayed how 600 RAF bombers attacked Penemunde on August 17, 1943. This raid supposedly delayed the V-weapon attacks for six months and perhaps even helped to secure the success of the Allied invasion on D-Day.[134] But the Allied scientists had been wrong before, especially when claiming in June of 1944 that there were no German V-weapons beyond the V-2, since a Vergeltungswaffe-2 would require a range of 100 miles—a claim so fantastic that "even now [June 1944] many military experts are skeptical."[136]

In the midst of the V-attacks, the New York *Times*, which claimed Londoners "were constantly dodging bombs, crowding into shelters or darting under desks and tables," also relayed a story of one American who "vowed he would get his twenty-eight German planes before he flew home to Oil City, Pennsylvania, to marry his girl."[137] And British newspapers, which were censored and forbidden to carry published attack information or publish photographs were allowed to give correct casualty figures in order to lift public morale.[138] Such stories were designed to keep the German uninformed about V-weapon success, and, at the same time, diminish Londoner concern over casualties.

However, the Germans needed little by way of British encouragement, since statements like the following from Voelkischer Beobackter were in ample supply:

> It must be remembered by one and all should they fear they are faltering, that technological superiority alone wins wars, just as ideas in the final analysis triumph in politics.[139]

The German claim that the V-weapons were superior technologically was rejected by the following account which appeared in the New York *Herald Tribune*. It seems that V-1 carried inferior explosives as compared to the Allies RDX (which was forty percent more explosive than TNT)—which were largely calcium nitrate. The story quickly pointed out that calcium nitrate was little more than chemical fertilizer. The whole discussion led one correspondent to remark: "Certainly, I for one would rather have a load of fertilizer dumped on me than a flying bomb."[140]

vii. Life Goes On

By September of 1944 the blackouts were lifted and the V-attacks were virtually ended. Supposedly, Londoners were more cheered by the lifting of the blackout than by the apparent end of the attacks since the "blackouts were termed the 'chief' wartime cause of crime and accidents to civilians."[141] In effect, while new and frightening technological weapons of World War II did indeed protract the night, they did so only temporarily. Culture may have slowed but it did not stop since few cultural activities, popular or otherwise, are long suppressed by technology (fig. 32-35).

Fig. 32
Variety, Oct. 23, 1944, p. 2.

Fig. 33
Variety, Oct. 23, 1944, p. 3.

Wednesday, November 1, 1944 VARIETY 31

KEEP YOUR EYES ON

WNEW

TELEVISION

America's First Independent Station to Televise Regularly!

WHAT does post-war television hold for the country's independent stations?

Can they stand the bill? Supply the talent, programs, and showmanship to compete with the larger outlets?

We, for one, are not going to sit around while someone finds out for us. WNEW wants the answer now!

That's why we are televising *now*—programs from our regular schedule we think will make good *visual* entertainment. You can see them on the DuMont Television Station WABD—Channel 4 on your receiver.

More important, our own writers prepare the shooting scripts—our own staff produces, directs, and handles the cameras. We want *them* to know what's going on. They will contribute much to our television future.

Sure it's an experiment. But experiment brought forth 24 hour-a-day broadcasting, the Make-Believe Ballroom, the Milkman's Matinee, and many other features that have made WNEW the most listened-to non-network station in the country. So keep your *eyes* on America's first independent station to televise regularly—WNEW!

WNEW

501 MADISON AVENUE, NEW YORK 22, NEW YORK

Fig. 34
Variety, Nov. 1, 1944, p. 31.

Wednesday, July 18, 1945

The artist gives his impression of Margaret O'Brien and Jackie "Butch" Jenkins from the M-G-M true-to-life drama "Our Vines Have Tender Grapes."

LIKE A DIAMOND IN THE SKY!

Something to sing about is M-G-M's "OUR VINES HAVE TENDER GRAPES."

We screened it this week with joy in our hearts!

Friendly customers! Watch for the Trade Show so that you will know how wonderful it is.

Thank you Edward G. Robinson for your stellar performance. Bless you little Margaret O'Brien for yours. And Butch, you are terrific!

Yes, the BIG ONES come from M-G-M!

EDWARD G. ROBINSON • MARGARET O'BRIEN in "OUR VINES HAVE TENDER GRAPES" with

Fig. 35
Variety, July 18, 1945, p. 20.

Grin and Bear It **By Lichty**

"The war sure aged him something awful—he says comic books leave him cold!"

Fig. 36
Washington *Post,* July 30, 1945, p. 10.

People were wondering what Charlie McCarthy would look like on television.[142] The *Saturday Review* carried photographs of the best U.S. novelists. Based on contributors' votes, the following list, in descending order, was completed: Ernest Hemingway, Willa Cather, John Dos Passos, Sinclair Lewis, Thomas Wolfe, Ellen Glasgow, Theodore Dreiser and John Steinbeck.[143] Cecil B. DeMille's "The Story of Dr. Wassell," starring Gary Cooper as a missionary in war, was featured at theatres across the country. Museum attendance dropped after Pearl Harbor but picked up again, according to "Interest in Art Persists in Spite of War's Demands."[144] Silver sold at 71.7 cents an ounce on September 16, 1942 and Babe Zaharias won the medal in Western Golf. Joseph Pulitzer, St. Louis publisher, was indicted by a Federal Grand Jury on three counts of game law violations from an "illegal shipment of ducks from Arkansas."[145] A beekeeper applied for a patent for a separation through which bees from different hives would chew at different rates in order to become acquainted, thereby not meeting en masse and destroying each other's swarms. People heard Bruno Walter at the Philharmonic, saw *Fantasia* at the Broadway, *Gone With The Wind* at the Capitol, *Santa Fe Trail* at the Plaza and Paul Robeson in Eugene O'Neill's *Emperor Jones* at the Fifth Avenue Playhouse. Happy husbands drank Beech Nut Coffee, Fordham beat Purdue in football 17-0, and concert goers had a choice between Claudio Arrau, Eugene Ormandy, Leopold Stokowski, Marian Anderson, Arthur Rubinstein or Isaac Stern.

The weather forecast for Chicago, June 16, 1944: Partly cloudy, continued warm and humid, high today 90, low tonight 68; Sunrise at 5:15 a.m.[146] For a time, at least, there were no longer to be protracted nights ruled by the mysterious horrors of war.

Notes

[1]Tomlinson, H.M., "Frankenstein in England," *Atlantic Monthly*, Oct. 1944, 174:57.

[2]The V2 and radar were truly secret and each surprised the respective enemy.

[3]"Novel Patents of the Week," New York *Times*, 14 Dec. 1941, II, 7:5.

[4]*Ibid.*, August 17, 11 5:4, 1941.

[5]"Unusual Patents of the Week," New York *Times*, Dec. 14, 1941.

[6]"Japanese Forces, (et. al.)," New York *Times*, Dec. 14, 1941, front page.

[7]"Cosmic Rays Tell Plane Altitude," New York *Times*, Dec. 14, 7:5.

[8]*Ibid.*

[9]New York *Times*, Dec. 12, 1941.

[10]"Emergency Rations Found in Downed Japanese Plane," New York *Times*, Dec. 28, 1941, p. 12.

[11]"Pilot Suits With Heat," New York *Times*,Nov. 2, X 3:4, 1941.

[12]*Ibid.* [13]*Ibid., "Along The World's Far Flung Airways."*

[14]*Ibid.* [15]*Ibid.*

[16]"Airplanes and Noise," New York *Times*, March 30, X 4:5, 1941.

[17]"Naval Officers' Device to Help Blind Flying," New York *Times*, April 20, II 7:4, 1941.
[18]*Ibid.*
[19]*Newsweek*, April 20, 1942, pp. 50-51.
[20]New York *Times*, Sept. 20, 1942, p. 13.
[21]*Ibid.*
[22]"Instruments," New York *Times*, July 26, IV 7:7, 1942.
[23]"Another Bombsight is Patented," New York *Times*, Oct. 4, III, 1:2, 1942.
[24]"Develops Plane Finish," New York *Times*, August 19, 31:5, 1942.
[25]"Post-War Airways May Use Radio Beams in Triplicate," New York *Times*, July 26, II 12:4,5, 1942.
[26]"Louis Signs at $21, Eager to Get Japs," New York *Times*, Jan. 13, 1942, p. 10.
[27]*Ibid.* [28]*Ibid.*
[29]New York *Times*,March 4, 1943, front page headlines.
[30]"Motors Do Many Jobs," New York *Times*, Jan. 13, 14 13:4, 1943.
[31]"New Fortress Deadly," New York *Times*, Jan. 3, Ix 41:8, 1943.
[32]"Planes Well Armed," New York *Times*, Jan. 3, IX 42:8, 1943.
[33]"Industry Has Part in Air Supremacy," New York *Times*, Feb. 7, 24:7, 1943.
[34]New York *Times*, Jan. 8, 1944, p. 22.
[35]"Flying Fortress Are the Special Pride of Seattle Workers," New York *Times*, Jan. 3, IX 43:1-5, 1943.
[36]"Changeover Is Made Without Time Loss," New York *Times*, Jan. 3, IX L4:8, 1943.
[37]"Arms Goal in 1943," New York *Times*, Jan. 3, 1943, Front page.
[38]"Girl Grows Curls to Help Win War," New York *Times*, May 7, 8:4, 1943.
[39]"Surgeon Honored for Flak-Proof Vest," New York *Times*, July 9, 6:2, 1943.
[40]"Flier Can Now Dictate Observations In The Air," New York *Times*, August 27, 2:4, 1943.
[41]"Fliers on Rafts Use 'Gibson Girl' To Send SOS, Guide Searchers," New York *Times*, May 17, 17:3, 1943.
[42]*Science News Review,* New York *Times*, April 11, IV 13 6, 7, 1943.
[43]"On Wings of Wood," New York *Times*, Sept. 12, IV 11:7 1943.
[44]"Army Reveals Antiaircraft Gun Good at 60,000 Feet, A 50% Gain," New York *Times*, June 24, 1:2, 1943.
[45]New York *Times*,June 24, 1943, p. 7.
[46]"The Advancing Art of War: Response to General G.M. Barnes," New York *Times*, Sept. 24, 1943.
[47]New York *Times*, Sept. 24, 1943.
[48]New York *Times*, Jan. 3, 1943, p. A13.
[49]"In Science Fretch," *Science News Letter*, Oct. 30, 1943, Vol. 44, No. 18.
[50]"General's Air Crash Stirs Suggestion," New York *Times*, Feb. 6, 27:4, 1943.
[51]"End of A Secret," *Newsweek*, April 26, 1943, p. 29.
[52]"Churchill Reveals Nazi Using Glider Bomb Powered by Rocket," New York *Times*, Sept. 22, 13:4, 1943.
[53]"Glider Rocker Bombs," New York *Times,* Sept. 24, 22:3, 1943.
[54]Cf. Fn. 1, page III, 1.
[55]"The New Weapon," New York *Times*, Sept. 26, IV 13:5, 1943.
[56]*Ibid.*
[57]"Secret Weapons I," New York *Times*, Sept. 30, 4:3, 1943.
[58]*Loc. cit.*
[59]"Nazi Rocket Weapon," *Science News Letter*, Oct. 30, 1943, Vol. III, No. 10, pp. 2-3.
[60]"Waiting for Robombs," *Science News Letter*, Vol. 47, No. 4, Jan. 27, 1944, p. 60.
[61]"Battle of England," *Time*, Vol. 44, Aug. 21, 1944, p. 29.
[62]"World Battlefronts: Battle of Europe," *Time*, Vol. 43, Jan. 3, 1944, p. 21.
[63]"Rockets," *Life*, Vol. 10, No. 3, Jan. 17, 1944, p. 71.
[64]*Ibid.*, p. 75.
[65]"Britons Prefer Bombings to New Mystery Plane," Ann Arbor *News*, June 16, 1944, front page.
[66]*Loc. cit.*
[67]New York *Times*, June 17, 12:3, 1944.
[68]*Ibid.* [69]*Ibid.*
[70]"Robot Raiders Attack Britain 2D Night in Row: Building Smashed by Weird Craft," Chicago *Daily Tribune*, June 17, 1944, front page.

[71]McKenzie, Donald, "Watches Robot Streak Skies and Blow Up," Chicago *Daily Tribune*, June 17, 1944, p. 6.
[72]*Ibid.*, July 18, 1944.
[73]"Two-Day Tryout Indicates Der Fuehrer's Latest Is No Wonder Weapon," Detroit *Free Press*, June 17, 1944, front page.
[74]"Planes With No Pilots," London *Times*, June 17, 1944, p. 4.
[75]*Ibid.*, p. 5.
[76]"Nazis Send More Robot Bombs to Bomb Britain," Washington *Post*, June 17, 1944.
[77]*Ibid.*
[78]"As We See It—Just a Nuisance," Detroit *Free Press*, June 17, 1944, p. 4.
[79]"War Games Summarized," New York *Times*, June 18, 1944, front page.
[80]"Britain Perfecting Antidote as Robot Bomb Plague Increases," New York *Times*, June 18, 1:5, 1944.
[81]"Coastal Guns Explode Nazis Robot Bombs in Air," Detroit *Free Press*, June 18, 1944, front page.
[82]"Nazi Propaganda Sears With Rocket," New York *Times*, June 18, 1944, p. 26.
[83]Avery, Marjorie, "Nazis Losing Again in 'Nerve War'," Detroit *Free Press*, June 19, 1944.
[84]"German People Deluded Fantasies of the Flying Bomb," New York *Times*, June 20, 1944, p. 3.
[85]"The Secret Weapons," *Newsweek*, Vol. XXIII, No. 26, June 26, 1944, page 24.
[86]Cromie, Robert, "Nazis Use New Type of Bomb to Hit Britain," *Chicago Daily Tribune*, June 22, 1944, Front page.
[87]*Ibid*, page 2.
[88]"New Robot Speeds at 600 Miles An Hour," *New York Times*, June 20, 1:7, 1944.
[89]Fuller, "Buzz Bomber," *Newsweek*, (Vol. XVIII, No. 26) June 26, 1944, page 24.
[90]*New York Times*, June 19, 1944.
[91]"International Control Asked," *New York Times*, June 28, 5:5, 1944.
[92]"Nine Kinds of V-1," *Newsweek* (Vol. XXIV, No. 7) August 14, 1944, page 34.
[93]"Foreign Affairs," *Newsweek*, (Vol. XXIV, No. 9), August 28, 1944, page 48.
[94]"Million London Homes Blasted in Robot Blitz," *Chicago Daily Tribune*, September 16, 1944, page 8.
[95]Young, Michael, "Life Under the Robot Bombs," *The New Republic*, (Vol. III, No. 10), September 4, 1944, pages 271-272.
[96]"Battle of England, The Blitz and One Man," *Time*, (Vol. 44) August 21, 1944, pages 27-28.
[97]"Correspondence," *The New Republic* (Volume III, No. 8), August 21, 1944.
[98]"Away From the Blastards," *Newsweek* (Vol. XXIV, No. 4), July 24, 1944, page 33.
[99]"V-2 Blasts and Shakes England," *Life*, (Vol. 17, No. 23) December 4, 1944, page 108.
[100]Bently, Phyllis, "A Report On V-1 and V-2," *Saturday Review*, February 10, 1945, page 17.
[101]*Ibid., Newsweek*, "Nine Kinds of V-1," page 35.
[102]"Battle of Britain Receiving End," *Time*, August 7, 1944, page 26.
[103]"Robot Blitz Kindling Hatred Instead of Fear in Britain," *Newsweek*, (Vol. XXIV, No. 9), August 28, 1944, page 48.
[104]*Ibid, Time*, August 7, 1944, page 26.
[105]"The Enemy: V-2," *Time*, (44 No. 21), November 20, 1944, page 28.
[106]H.M. Tomlinson, "Frankenstein in England," *Atlantic Monthly*, October 1944, page 59.
[107]Saunders, Hillary St. George, "The Flying Bomb," *Life*, (Vol. 17) November 20, 1944.
[108]*Ibid.*, Tomlinson
[109]*Ibid.*, Saunders.
[110]"The Devil's Workshop," *Newsweek*, October 2, 1944, page 32.
[111]"The Buzz Bomb Dooms Germany," *Saturday Evening Post* (217), September 1944, pages 14-15.
[112]*Ibid*, "The Enemy V-2," *Time*, page 28.
[113]"Battle of Britain: Epilogue." *Time*, October 23, 1944, pages 31-33.
[114]"Resume Robot Raids on Eve of Lights Return," *Chicago Daily Tribune, September 17, 1944, page 4.*
[115]"Robot Planes Not New," *Science News Letter*, (Vol. 46, No. 2), July 8, 1944.
[116]*Ibid.*
[117]"Secret of Robot Bombs May Be In Patent," *Science News Letter*, July 1, 1944.

[118]Hemingway, Ernest., "London Fights the Robots," *Collies,* August 19, 1944.
[119]"Battle of England: The Score for Robots," *Time,* September 18, 1944, page 35.
[120]"U.S. Invention Helps Shoot Down Robot Bombs," *Newsweek,* December 24, 1944, page 49.
[121]*Loc. cit., Time,* September 18, 1944.
[122]*Ibid.*
[123]Summers, Martin, "The Buzz Bombs Will Revolutionize Warfare," *Saturday Evening Post,* September, 1944.
[124]"V-2 Weapons May Win Next War, But Come Too Late For This One," *Newsweek,* (Vol XXIV, No. 21), November 20, 1944, page 39.
[125]*Ibid.*
[125]*Ibid.*
[126]"Robot Bomb," *Life,* (Vol. 17 No. 2), July 10, 1944, pages 35-36.
[127]"The Buzz Bomb Dooms Germany," *Saturday Evening Post*, Sept. 1944, p. 14.
[128]Fuller, J.F.C., "The Flying Bomb Points the Way to the War of the Future,"*Newsweek,* July 17, 1944.
[129]*Ibid.*
[130]Fuller, et. al., "The Future of War VI," *Newsweek*, Dec. 25, 1944, p. 24.
[131]Haldane, J.B.S., "Science: V-3," *Time*, Vol. 44, No. 22.
[132]"Robomb Blasts Cause Plant to Fold Its Leaves," *Science News Letter*, Vol. 46, No. 25, Dec. 16, 1944, p. 392.
[133]*Ibid.*
[134]Miche, Allan A., "Forty Minutes That Changed the War," *Reader's Digest,* Oct. 1944, pp. 18-20.
[135]"V-2: Nazi Rocket Details Finally Revealed," *Life*, Vol. 17, No. 26, Dec. 25, 1944, p. 47.
[136]"Britons Steeled For Bigger Rocket Attacks," New York *Times*, 5:23k June 30, 1944.
[137]"Vergeltungswaffe Zwie," *Newsweek*, Vol. 24, July 17, 1944, pp. 20-21.
[138]"British Ask Data on Robot Deaths," New York *Times*, June 30, 1944.
[139]"New Arms Promised to Impatient Reich," New York *Times*, June 30, 1944, 4:4.
[140]*Newsweek*, Vol. 24, NO. 8, August 21, 1944, p. 31.
[141]"Lights On," *Newsweek*, Vol. XXILV, No. 12, Sept. 18, 1944, p. 31.
[142]"Imagine Charlie McCarthy on Television," *Life*, Oct. 23, 1944, p. 101.
[143]"Miscellaneous Photos of Best U.S. Novelists," *Saturday Review*, August 14, 1944, p. 14.
[144]"Interest in Art Persists," New York *Times*, Jan. 1942, II 7:5.
[145]"Joseph Pulitzer Indicted," New York *Times*, Sept. 17, 1943, II:7.
[146]Chicago *Daily Tribune*, June 16, 1944, front page.

Chapter Three
The Atomic Bombing of Hiroshima

Upon announcing the atomic bomb attack on Hiroshima, August 6, 1945, President Harry Truman, aboard the cruiser *Augusta* steaming homeward from the Potsdam meeting, announced to an assembled crew:

"We have just dropped a bomb on Japan which has more power than 20,000 tons of TNT. It was an overwhelming success." The applause and cheering hardly died down before the president left.[1]

This official announcement of what many term the Birth of the Atomic Age was the first of uncounted reactions to what has been considered either a scientific wonder or a scientific blunder. Most reactions were, and continue to be, of the unofficial variety, and the responses vary from being extremely precise scientifically to being extraordinarily emotional. Today, of course, one would expect immense political, social, military and personal reactions to such a massive technological intrusion into the order of things; however, a world weary of war, accustomed to death and concerned about shortages of cooking oil found itself bewildered and confused, hopeful and fearful, confident about the future but deadly afraid to leave the present. Since the press of the times carried some of the reactions and most of the information, the readers of newspapers received ample grounds upon which to make at least quasi-accurate appraisals of those atomic events surrounding the last few days of the war in the Pacific.

In order to provide some insight into the interplay of the press and the times, three perspectives are helpful: some of what the readers were exposed to before the bombing, the immediate reporting of the atomic attack, and the delayed evaluations of 1945.

i. Prelude to Disaster: Massive Bombings, Casualty Lists and Nylons by New Year's

The readers of the New York *Times* in the days preceding the atomic bombing learned that the Rock Island Line not only provided the finest in rail transportation but it also built the first bridge to cross the Mississippi River, or at least claimed an advertisement sandwiched between a notice that a mutual fund would pay three percent (insured up to $5,000) and a spot indicating that family grave plots (for up to four members) cost only $125. Readers were told that Japan had no battleships left, that the new P-80 jet plane could exceed 500 miles per hour, that twenty giant DC-6s (fifty-six passengers) had been ordered, that sport jackets cost $26.50 and slacks-to-match cost only $10.50. One could wear his new outfit to the Cafe Society and hear Mary Lou Williams (featuring jazz pianist Imogene Coca, fig. 37). Cab Calloway was at the Zanzibar (all for the price of the dinner alone), and at Rockefeller Center Gene Tierney, John Hodiak and William Bendix starred in *A Bell for Adono*.[2] Ninety percent of Southern California restaurants would be forced to close within two weeks because meat rations had been cut twenty percent, thereby forcing a stoppage of meals for 1,500,000 war workers in the area.[3] Hal Newhouse defeated the Browns 5-4 in a Detroit baseball victory. Gas stoves were taken off the ration list. On a more international level, the text for "an education and cultural organization of the United Nations" was being prepared for the allied ministers of education.[4]

In Washington, D.C., plastic handbags sold for five dollars each, plus a twenty percent tax. The July 1, 1945 Washington *Post* told that 45,000 planes were to be declared surplus by the year's end[5] and that Dick Powell and June Allyson had announced plans to be married soon.[6]

Newspapers across the country all carried the following in much the same way: "1,000 Planes Raid Tokyo Area"; "Army Will Use 8000 Planes in All Out Attack on Japan."[7] "1,500 or More American and British Planes Set Ablaze a 400-mile Strip of Japan's Coast";[8] "Record B-29 Attacks Unload 6,632 Tons of Bombs on Japs";[9] and, "Quick Reconversion Would Bring Some Economic Profit, Says City National Bank" if the war could be ended quickly.[10] One editorial article by Alexander F. Jones warned that within ninety days "every city of consequence in the home islands will be in shambles, its oil and steel destroyed, its food supply decimated" from the experience gained by the American Air Force during campaigns in Europe and Africa.[11] Throughout the pre-bombing period there is virtually no mention of anything atomic—it was indeed the best kept secret of the war.

Attempts were made to report on the less burdensome aspects of the Pacific War. Special wireless to the New York *Times* gave forth the news that a "Lilliputian Invasion Wins Philippine Isle" when the "smallest invasion of the war took place on Fuga

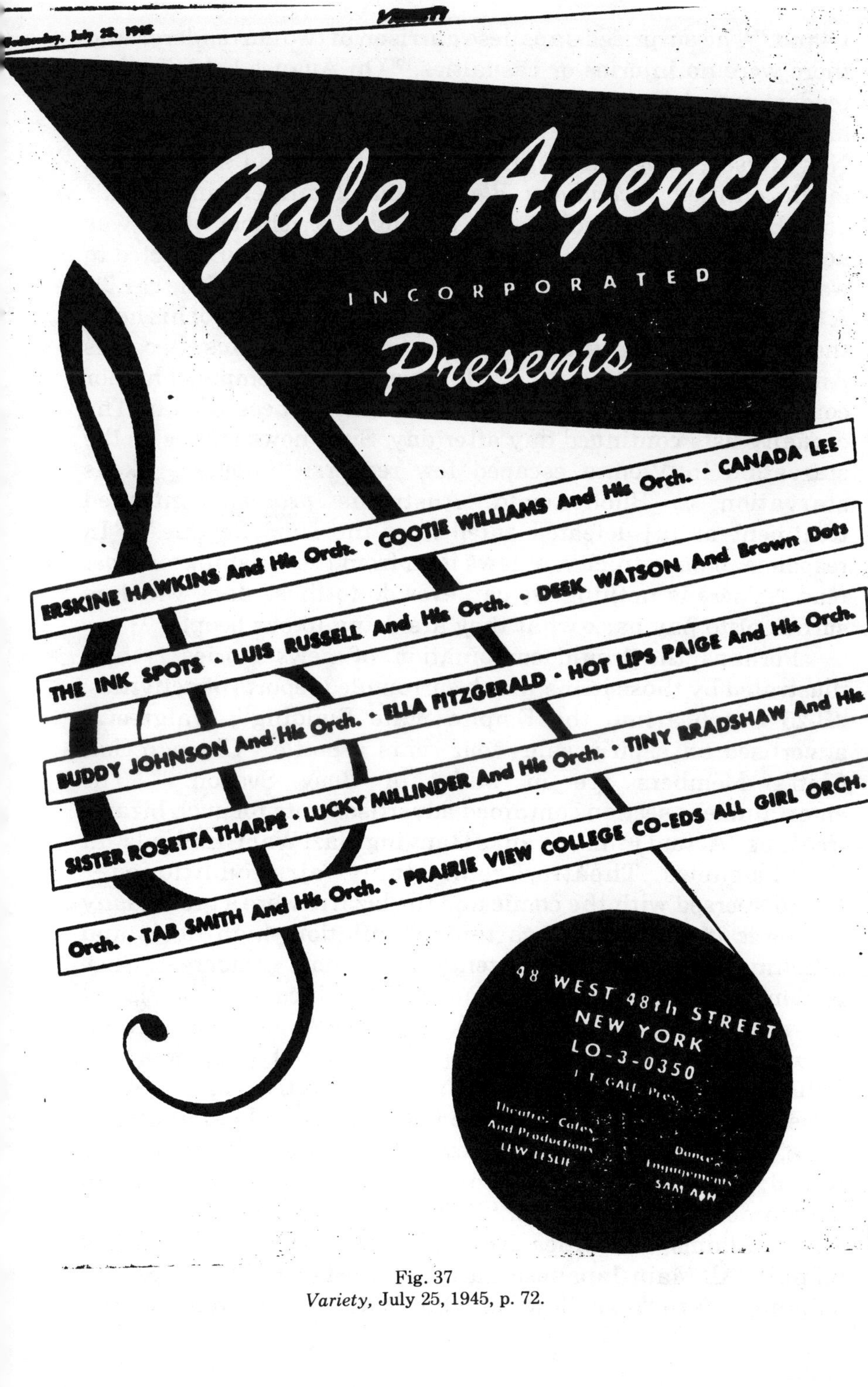

Fig. 37
Variety, July 25, 1945, p. 72.

Island." The surprised Japanese garrison of two surrendered and there were no injuries or casualties.[12] On August 1, 1945, some readers may have found some humor in the U.S. Army's latest announced secret weapon: DDT for the war against mosquitoes in New Jersey, "beginning Friday."[13] Betty Hutton starred in a movie entitled *Incendiary Blonde*—pun definitely intended.[14] Jack Benny, the comedian, was driving in a vehicle which was accidentally fired upon: "the sentry, who had been instructed to watch for marauders, fired at a tire but hit the body of the car."[15] A buffalo rancher in Humboldt, Iowa, slaughtered one of his herd, thereby prompting this headline: "Buffalo Au Jus Becomes Another Vanished Species."[16] However, such attempts at humor could detract only momentarily from the sadness of war. The casualty lists continued day after day. Such news stories as the starvation in Vienna escaped few readers: "Populace likens starvation to Buchenwald. Austrians protest continued treatment as [a] defeated satellite of the Nazi Regime."[17] In response to one atrocity, a news item from the Washington *Post* read: "There is nothing we can ever do to those Japs which is sufficient to pay back what they are doing to our people."[18]

Perhaps the ironic combination of news stories is best illustrated by those items which surrounded reports of activities: "B-25 Crashes into the Empire State Building"; Walgreen's advertised an aspirin sale—four cents a bottle: "Chicago Golf Club's Members are in Midst of Busy Season."[19] The entertainment section contained advertisements for such bizarre treats as "Actual Film of Yanks Hanging Nazi War Criminals" at the Telenews Theatre, comfortably air conditioned.[20] Interdispersed with the comic and the bizarre were a great many economic items which in a time of inflation, high taxes and rationing attracted the attention of most readers. "Shoe Rationing to End by Next January,"[21] "Chicago Girls Rise at Dawn to Detassel Corn" (in order to hasten crop ripening),[22] "Army Seizes Ward's Fashion Department for 2D [sic] Time" (for failure to place national war labor board directives into effect),[23] "Cost of Living Rose 29.9% Since 1940" (fruits and vegetables up 125%, meat 34% and dairy products 27%),[24] and similar reports are found scattered throughout omnipresent and often overburdening war reports: "D-Day Expected Within 60 Days, Tokyo Claims,"[25] "Cities Next on B-29 List" (with attached map),[26] "All Main Japanese Harbors Mined by B-29s" (Japanese shipping cut to "a shadow" of what it was earlier in the war).[27]

Nor could the reader escape from the pen of the propagandist:

Japan On The Spot

Now that Suzuki has rejected our terms the bombing should be stepped up, if that is possible. We can show the Japanese that the Potsdam terms are irreducible, that this Nation is inflexible in support of our fighting forces in the subjugation of Japan. That the war against Japan has become plain murder has become entirely Japan's fault.[28]

In the midst of such rancor and hatred one finds hope for the human condition in a short item virtually hidden on page 36 of the New York *Times*: "Race talks planned by Mrs. Roosevelt"—an adult workshop on inter-racial relations to be conducted at a place known as the Downtown Community School—"Other lecturers will include Mrs. Paul Robeson."[29]

ii. Before the Smoke and Dust Could Clear[30]

"Atomic Bomb Packs Punch Equivalent to 2000 B-29s" screamed the headlines of the Ann Arbor *News*.[31] The Associated Press reported: "Atomic Bomb U.S. Pulverizes Japan with Bomb Greater than 20,000 Tons of TNT,"[32] "Tokyo May be Next Target for Explosives."[33] "Power of Sun Harnessed to Make New Bomb."[34] Such headlined stories appeared in virtually every American daily newspaper on either August 6 or August 7, 1945 (figs. 38, 39, 40). Usually the afternoon papers carried the August 6 headlines; the morning editions could not report the atomic attack until Tuesday, August 7.

The first reports were generally characterized by an emphatic statement of awe and wonder at such massive technological success. Readers were told that science and industry had worked together (with Army supervision) to accomplish the "greatest achievement of organized science in history."[35] This marvel was the "achievement of scientific brains putting together infinitely complex pieces of knowledge" in numerous and varied sciences.[36] This secret of atomic energy had been "wrest from the universe" by 5000 scientists who worked at the University of Chicago for three years.[37] Such efforts were acclaimed as "mankind's successful transition to a new age" and as a success "greater than the most ambitious estimates."[38] One editorial read: "The discovery is the greatest scientific achievement of this war and it may be the greatest in history."[39]

The Weather
Today—Mostly sunny and warm but not so humid. Highest temperature in the middle 80s. Chance for showers by evening. Yesterday—High, 88; low, 78. (Details on Page 10.)

The Washington Post

Single Copy Price 5c

NO. 25,254 — WASHINGTON: TUESDAY, AUGUST 7, 1945

Single Atomic Bomb Shakes Japan With Force Mightier Than 20,000 Tons Of TNT to Launch New Era of Power

Prelude to Electronic Age Ushered in on July 16 Before Tense Group of Scientists at New Mexico Proving Grounds

Aware of the historic consequence of the first experimental detonation of the atomic bomb the Army carefully recorded the drama of the occasion. Here is the Army's story.

Mankind's successful transition to a new age, the Atomic Age, was ushered in July 16, 1945, before the eyes of a tense group of renowned scientists and military men gathered in the desertlands of New Mexico to witness the first-hand results of their two billion dollar effort.

Here in a remote section of the Alamogordo Air Base 120 miles southeast of Albuquerque the first man-made atomic explosion, the outstanding achievement of nuclear science, was achieved at 5:30 a. m. of that day. Darkening heavens pouring forth rain and lightning immediately up to the zero hour heightened the drama.

Mounted on a steel tower, a revolutionary weapon destined to change war as we know it or which may even be the instrumentality to end all major wars, was set off with an impact which signalized man's entrance into a new physical world. Success was greater than the most ambitious estimates.

A small amount of matter, the product of a chain of huge specially constructed industrial plants, was made to release the energy of the universe locked up within the atom from the beginning of time. A fabulous achievement had been reached. Speculative theory, barely established in prewar laboratories, had been projected into practicality.

This phase of the Atomic Bomb Project, which is headed by Maj. Gen. Leslie R. Groves, was under the direction of Dr. J. R. Oppenheimer, theoretical physicist of the University of California. He is to be credited with achieving the implementation of atomic energy for military purposes.

Tension before the actual detonation was at a tremendous pitch. Failure was an ever present possibility. Too great a success, envisioned by some of those present, might have meant an uncontrolled, unusable weapon.

Final assembly of the atomic ...

Atomic Bomb Packs Punch Of 2000 B-29s

Principle of the Atomic Bomb in Diagram

B-29s Blast Jap Naval Arsenal

Guam, (Tuesday), Aug. 7 (AP)—Approximately 125 B-29s struck the Tokyokawa naval arsenal with high explosive bombs around noon today.

There was no mention of the atom-bombing of Hiroshima. Shortly before the communique was issued, Spaatz's headquarters said all Hiroshima eyewitness stories would be cleared through the War Department, Washington.

Guam (Tuesday), Aug. 7 (U.P.)—More than 400 fighters and bombers of Gen. George C. Kenney's Okinawa-based Far Eastern Air Force demolished a Japanese "mystery town" Sunday in what possibly was an attack on newly-developed Japanese rocket-launching installations, front dispatches reported today.

Liberators, Mitchells, Invaders, Mustangs and Thunderbolts heaped 370 tons of gasoline-jelly bombs on the southern Kyushu city of Tarumizu, also known as Tarumi, and left it a sea of roaring flames. Returning pilots told of seeing a huge catapult-like machine extending over the water. It appeared to be similar to the rocket-launching devices used by the Germans in their bombardment of Britain.

Tokyo Area Blasted

Fleets of Iwo-based Mustang fighters, following in the wake of a giant predawn Superfortress assault which left fires visible for 150 miles in four enemy "death list" cities, carried forward the air assault on the enemy homeland by daylight yesterday, blasting the Tokyo area and destroying or damaging 23 enemy planes.

Tarumizu, an aircraft center, is ...

President Proclaims First Use of Greatest War Secret; Predicts Day When Gigantic Energy Will Change Industry

By Edward T. Folliard

Mankind has entered a new age, the age of atomic energy.

President Truman proclaimed the world-shaking event yesterday in a White House statement announcing the first use of history's greatest secret weapon—a single bomb that 16 hours before had dropped on Hiroshima, Japan, with an earthquake force greater than 20,000 tons of T.N.T.

"It is an atomic bomb," said the White House statement. "It is a harnessing of the basic power of the universe. The force from which the sun draws its power has been loosed against those who brought war to the Far East."

The explosion of that bomb, the most terrible missile ever invented, a fantasy at last fulfilled, must have been felt the length and breadth of Japan. Three weeks ago, when it was first tried out in the New Mexico desert, it was heard 300 miles away. In Albuquerque, 120 miles away, the flash caused a blind girl to exclaim "What was that?" The steel tower from which the bomb was dropped was transformed into vapor.

The bomb certainly will shorten the war against Japan, if it does not end it. And in the language of the War Department, it "may even be the instrumentality to end all wars."

Hoped to Spare Japs

President Truman, in a statement wirelessed from the cruiser Augusta, said it was to spare the Japanese people from "utter destruction" that the Potsdam ultimatum had been issued on July 26. Then, in what amounted to a fresh ultimatum, the Chief Executive said that if the Japanese do not now accept the Allied terms "they may expect a rain of ruin from the air, the like of which has never been seen on this earth."

For the moment, interest centered on the military potentialities of the great discovery. But the American and British scientists who had developed it saw it as a power for good as well as for evil, ...

Churchill Hails Roosevelt Role In Discovery

London, Aug. 6 (AP)—Former Prime Minister Churchill said tonight that it was "by God's mercy" that American and British, instead of German scientists, discovered the secret of atomic power "long mercifully withheld from man."

The success of the historic achievement, he added, stood "to the everlasting honor" of the late ...

Fig. 38

"All the News That's Fit to Print"

The New York Times.

LATE CITY EDITION

VOL. XCIV..No. 31,972. NEW YORK, TUESDAY, AUGUST 7, 1945. THREE CENTS

FIRST ATOMIC BOMB DROPPED ON JAPAN; MISSILE IS EQUAL TO 20,000 TONS OF TNT; TRUMAN WARNS FOE OF A 'RAIN OF RUIN'

HIRAM W. JOHNSON, REPUBLICAN DEAN IN THE SENATE, DIES

Isolationist Helped Prevent U.S. Entry Into League—Opposed World Charter

CALIFORNIA EX-GOVERNOR

Ran for Vice President With Theodore Roosevelt in '12—In Washington Since '17

Jet Plane Explosion Kills Major Bong, Top U.S. Ace

Flier Who Downed 40 Japanese Craft, Sent Home to Be 'Safe,' Was Flying New 'Shooting Star' as a Test Pilot

KYUSHU CITY RAZED

Kenney's Planes Blast Tarumizu in Record Blow From Okinawa

ROCKET SITE IS SEEN

125 B-29's Hit Japan's Toyokawa Naval Arsenal In Demolition Strike

REPORT BY BRITAIN

'By God's Mercy' We Beat Nazis to Bomb, Churchill Says

ROOSEVELT AID CITED

Raiders Wrecked Norse Laboratory in Race for Key to Victory

Steel Tower 'Vaporized' In Trial of Mighty Bomb

Scientists Awe-Struck as Blinding Flash Lighted New Mexico Desert and Great Cloud Bore 40,000 Feet Into Sky

'NEW AGE USHERED

Day of Atomic Energy Hailed by President, Revealing Weapon

HIROSHIMA IS TARGET

'Impenetrable' Cloud of Dust Hides City After Single Bomb Strikes

MORRIS IS ACCUSED OF 'TAKING A WALK'

CHINESE WIN MORE OF 'INVASION COAST'

ATOM BOMBS MADE IN 3 HIDDEN 'CITIES'

TRAINS CANCELED IN STRICKEN AREA

Fig. 39

The Weather
Today—Pleasant, sunny weather with moderate daytime temperatures; highest near 80 and low humidity. Cool tonight, lowest temperature near 60. Yesterday—High, 85; low, 65. (Details on Page 16.)

The Washington Post

News Every Hour On the Hour
Direct from our News Room at The Washington Post
WINX
1340 On Your Dial

NO. 25,255 WASHINGTON: WEDNESDAY, AUGUST 8, 1945

Single Atomic Bomb 'Dissolved' Jap City Of 318,000, Returning U. S. Fliers Say

President Arrives At White House, Calls Cabinet

Advisers Will Meet On Friday; News Conference To Be Held After Radio Tal[k]

BULLETIN

Found Key To Atomic Energy

DR. LISE MEITNER,
Austrian-Jewish woman physicist driven from her homeland by Hitler's racial laws, who played an important part in the development of the atomic bomb by breaking a 10-year deadlock in experimentation. Her mathematical calculations proved her thesis that the splitting of uranium atoms into two parts released great stores of energy. Since leaving Germany in 1938 she has been at the Stockholm Academy of Science
(Story on Page 2)

New Weapon Causing No Cut In Army

Military To Play Safe In Keeping 7 Million Troops For Invasion Force

Area Struck by Atomic Bomb Is Saturated With Death for 70 Years, Scientist Reveals

By Dr. Harold Jacobson

War Climax Is Awaited In Capital

Quick Showdown With Japs Seen; New Warning By Truman Expected

By Edward T. Folliard

All Hiroshima Prefecture Is Isolated, Foe Asserts, As Accounts of Explosion Reach Hysterical Scream

Guam Aug. 8. (U.P.).—Four and one-tenth square miles of Hiroshima or 60 per cent of the city's built-up area were wiped out by a single atomic bomb Monday, it was officially announced tonight.

The area was "completely destroyed by the atomic bombing mission."

Five major industrial targets within this area were destroyed.

Additional damage was shown outside the completely destroyed area, Gen. Carl A. Spaatz announced in a Strategic Air Forces communique.

Guam, Aug. 8 (Wednesday) (INS).—The 12-square-mile city of Hiroshima with its 318,000 inhabitants "dissolved" in a vast cloud of dust when the world's first atomic bomb struck home on Japan shortly after 9:15 Monday morning (Tokyo time).

That was the only but cataclysmic result observed by the pioneering crew of the bomb that loosed the "capsule of devastation."

The bomb was officially described as having exploded with the destructive power of a combined normal bomb load from 2000 B-29 Superfortresses.

A trio of young American aviators chosen to launch the unprecedented attack summarized last night the effect as follows:

"The only way we could tell a city had been there was because we had seen it a moment before."

[When the bomb exploded, the United Press said, the men of the crew, 10 miles away, exclaimed as one: "My God!"

[What had been Hiroshima going about its business at 9:15 o'clock of a sunny morning, the United Press said, went up in a mountain of dust-filled smoke, black at the base, towering into a plume of white at 40,000 feet.—Editor's note.]

The only American eyewitnesses of the shattering attack by the new type bomb were permitted to tell their story as Tokyo broadcasters increased to a hysterical scream their own reports on the bomb's effect.

Entire Prefecture Isolated, Japs Report

Attesting to the devastation inflicted, the Tokyo radio said in a broadcast heard this morning in Guam that all communications with Hiroshima prefecture have been cut off.

First Major Action In Week

Carrier Planes Sweep China Coast In Three-Day Assault

Tuck Holding 3 to 1 Lead In Virginia Race

Fenwick Tops Three Man Contest; Vote Is Nearly Half In

News Photogs Mistook Mayflower

As immediate were the proclamations of military advantage. One writer editorialized that "Dropping of the first atomic bomb on the Japanese army base at Hiroshima was merely a warning to the enemy."[40] Another journalist suggested that the ultimatum of July 26 had been issued at Potsdam to spare the Japanese public from utter destruction,[41] and still another bragged that Japan had forty-eight hours to surrender or "face the prospect of the entire obliteration of the Japanese nation."[42]

Washington, D.C. August 6 (AP). Hiroshima, the first enemy city to feel the American atomic bomb was a major military target. Whether it is anything important now cannot be determined until the smoke and dust [have] cleared.[43]

"The tall balding navy officer told of his reaction . . . with the calmness of a professor lecturing to a group of students on butterflies," claimed a report in the Chicago *Tribune*. And the bombing was to continue until Japan surrendered for "her people would suffer a 'rain of ruin from the air' which would cause the population to perish in suffocation."[44] Some editors attributed the destruction caused by the bomb to the Japanese themselves: "The immediate future will show if our oriental enemy intends to commit national *hara-kiri*."[45]

"It is my firm belief," said Eddie Rickenbacker, "that the 'Japs'" will be demoralized by a "daily ration of atomic, demolition and incendiary bombs that they will cry quits."[46] *Newsweek* termed the atomic bomb a "wonder weapon" and claimed that the "Awesome force of the atomic bomb [was] loosed to hasten the Jap surrender."[47] *U.S. News* reported that no despotic mind could challenge the United States since the Americans had "the most potent weapon of the world" and since there were ample quantities in the U.S.[48] (fig. 41). "The War Ends—Burst of Atomic Bomb Brings Swift Surrender of Japanese" announced *Life* magazine.[49]

Before the smoke and dust could clear, however, the militarists were already suggesting that warfare would never be the same as it had been since ships, fortifications and pillboxes would disintegrate,[50] for a "revolution in science and warfare have occurred on the same day."[51] "War in the old sense is a thing of the past," claimed *U.S. News*, since "manpower and guns have become things of small moment."[52] *Newsweek* claimed that military confusion reigned as to what countermeasures could be used against the bomb; essentially, most experts agreed that

Fig. 41
Variety, August 23, 1945.

capturing the enemy's source of supply was the only effective defense.[54]

American readers also were able to read some of the foreign responses to the atomic attack. One United Press International story datelined August 6, 1945, claimed to have mentioned a Japanese broadcast from Tokyo which said the effects of the bomb would be nullified and termed America "public enemy No. 1 of social justice" for its attack on civilians. Radio Moscow broadcast the complete text of Truman's atomic statement "but made no comment."[55] The Federal Communication Commission noted that Domei had reported that a small number of B-29s had caused "considerable damage to Hiroshima" with a new type of weapon ... details are now under investigation."[56] "Vatican Paper Sorry Bomb Was Invented" and similar reports streamed from Rome, but there was no concrete confirmation of the Vatican's position.[57] Winston Churchill, not one to remain silent on historical events, praised the American and British effort and claimed it was "by God's mercy" that the Allies and not the Germans had "discovered the secret of the atomic bomb."[58] "Britons Revolted by Use of Atom Bomb" readers of the New York *Times* discovered. The commentaries appeared in a number of letters to the editors: "Is there to be no protest against the crime of Hiroshima?"; "Japan has never aroused my sympathy until today, and now my heart goes out to her"; "My God! Has the world gone crazy?"; "In the name of humanity, let us stop and ask where we are going"; and, "every instinct in humanity makes us recoil"[59] (fig. 42).

Japanese Radio Singapore announced that Japan has "a bomb similar to the atomic bomb,"[60] but the broadcast announced that there were no plans for the "deliberate massacre of innocent civilians" and therefore the bomb would not be deployed.[61]

> "To us the occurrence of this sort of weapon is nothing very novel," the broadcast said, "the theory of atomic energy has been known to us for the last century. The fact that the Anglo-American nations, who are far behind the times in this sphere of science, are putting to use the theory of atomic energy makes it clear they must have drawn on the results of Japanese researchers."[62]

The *Catholic World* said the bomb was an example of the "immorality along the path we have been treading" and the use of such a weapon was "utterly indefensible."[63] "The atomic bomb

Fig. 42
U.S. News, August 17, 1945, p. 21.

was for a people with strong superstitions Heaven's sign that God had foresaken them," commented the Turkish press.[65] Herman Goring, awaiting trial as a war criminal, said: "A mighty accomplishment. I don't want anything to do with it."[66] One columnist for the London *Sunday Express* wryly commented: "While the general election killed conservatism in England, the atomic bomb killed it all over the world."[67] *Newsweek* implied a foreign news story was a Japanese lie designed to gain international sympathy. In fact, the headline response to the comment that "many would not survive," was *Newsweek*'s "Pity Us, Thank You":

> The crowds of victims, their bodies burned red and black, seemed to him like a parade of ghosts from another world. Most of the injured . . . died in agony, and the living screamed at passers-by to kill them quickly.[68]

Despite such stories, there was an amazingly large number of "objective" reports, or at least reports which seemed to be so. One type was the "who-done-it" news story: "Workers Pay $1 a Paper to Learn About Product" read one headline about employees in Oak Ridge, Tennessee (where one vendor sold 1,600 newspapers in thirty-five minutes);[69] "There were strange looking people (who) conversed in strange languages" at the University of Chicago among the 5,000 scientists;[70] "this discovery was made by American scientists and engineers" in a free society;[71] "Oak Ridge Celebrates but Sticks to Bomb Job" so that after a night of celebrating "the absenteeism caused no loss in production";[72] and, "Truman and Secretary of State Byrnes deserved credit for the test as a whole, but Franklin D. Roosevelt deserved most credit for foresight and wisdom."[73]

Another type of "objective" reporting consisted of "how the bomb works" stories (fig. 43, 44). Virtually every newspaper carried information of the total impact value of the bomb (20,000 tons of TNT) and the total area destroyed (only 2.8 of 6.9 square miles of Hiroshima remained). The estimated heat from the bomb was placed at two trillion degrees centigrade and the American public read that it was "Greater in Fahrenheit (3,600,000,000,932^{0} F or 2 x 10^{12})."[74]

> It would be incorrect to say these victims burned to death because radioactive rays generated the incredible heat that explodes atoms. In the unimaginably short time when those human beings suffered the effects of the atomic bomb, there probably were millions of explosions touched off in the atoms which comprised the matter which was their flesh, blood, and bones.[75]

the power BEHIND atomic bomb production

The bulk of the steam used in meeting the vast power and process requirements of atomic bomb production is supplied by Combustion Engineering equipment.

The part played in this undertaking by any one organization is dwarfed by the breath-taking magnitude of the achievement itself. None the less, we are proud that equipment designed by our engineers and built by our shop men constitutes the principal source of steam power in this first successful process for releasing the energy dormant in the atom since the beginning of time.

May the very availability of atomic energy bring a quick end to World War II and serve as a strong deterrent to all future wars. And may it in peacetime fulfill its vast potential for constructive service to all mankind!

A 903

COMBUSTION ENGINEERING

200 MADISON AVENUE, NEW YORK 16, N. Y.

C-E INSTALLATIONS span the entire range of steam generating requirements from small stoker-fired boilers of less than 50 horsepower to the largest power station units.

Fig. 43
Business Week, Aug. 18, 1945, p. 27.

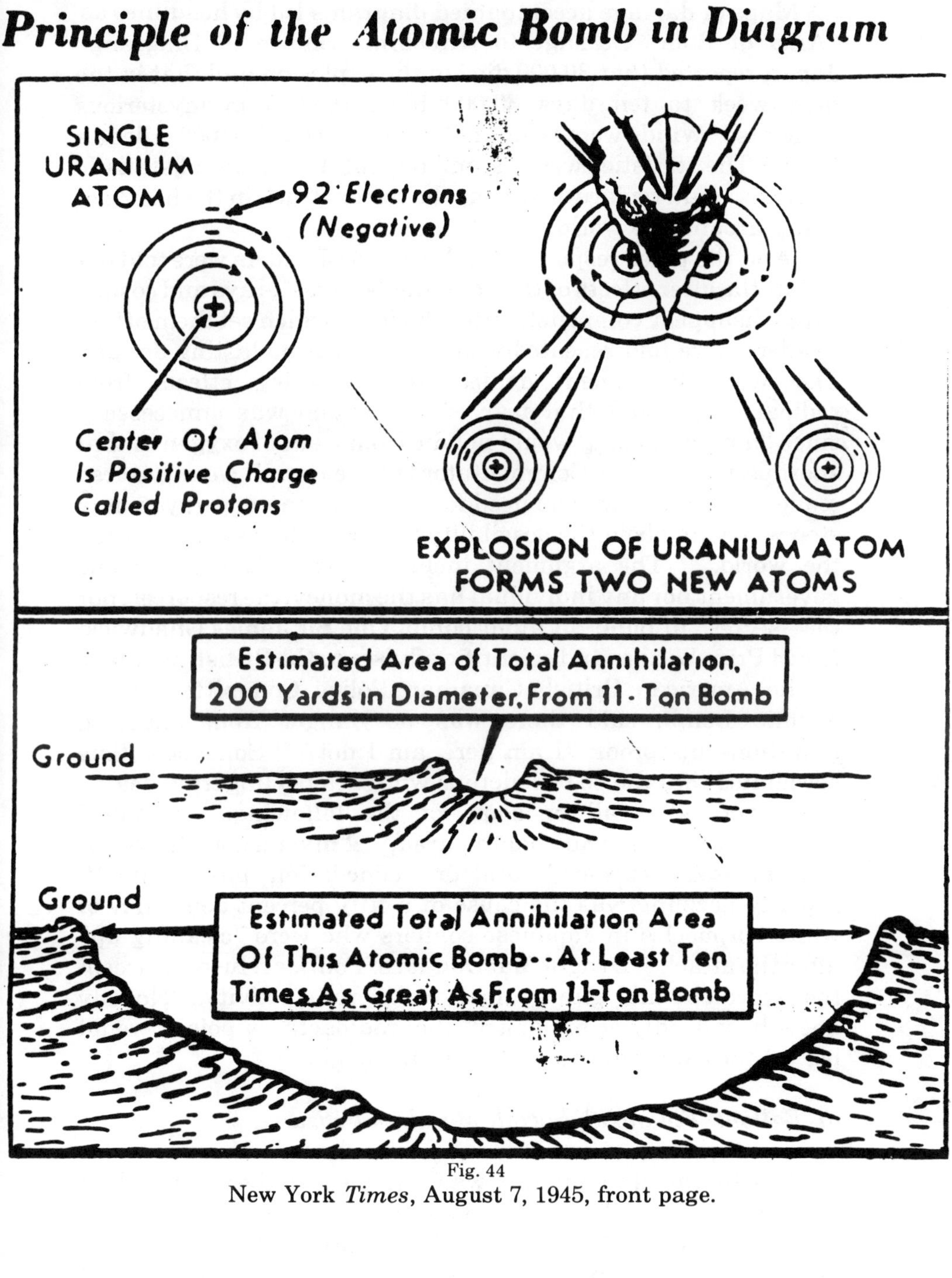

Fig. 44
New York *Times*, August 7, 1945, front page.

Maps of damage accompanied diagrams led by headlines as "Area of Bomb Damage to Hiroshima Shown in Diagram." Japan reported that 30,000 died in the explosion and that in the next week to ten days 30,000 more died from mysterious ailments.[76] Winds were estimated to be between 500-1000 miles an hour.[77] The statistics went on and on, but no amount of statistic analysis could replace or subdue the terror struck in the hearts of many around the world.

And the press responded to this terror. Readers were told not to fear the after effects of the bombing because x-rays and cosmic rays "disappear completely after the flash which set them off."[78] Readers were told repeatedly that there was no lasting danger. The War Department denied any lingering effects from radioactivity[79] and that fear of suffocation was unnecessary since "a human being would not die from want of oxygen in that short period."[80] "The Commentator" in the Detroit *News* assured readers that there was no need for concern: "moreover, you may stop worrying about the possibility that some looney will destroy the world."[80] The argument included the logic that no one government nor any individual has the money, the resources, nor the "secret" to build an atom bomb. One Sir James Chadwick, Nobel Prize Winner and scientific advisor to the British members of the American British Canadian Policy Committee on the Atomic Bomb, said there was no danger from lingering radiation—his proof: "I am here, am I not?"[81] Some scientists speculated that the 30,000 victims who suffered lingering deaths were victims of all or some of the following: "continued radioactivity in the bomb area," "mysterious rays of concentrated neutrons," and/or "concussion pneumonia."[82] Equally puzzling was the cause of a fifty percent deficiency of white corpuscles in Japanese soldiers who were "cleaning up" after the attack.[83] Dwight MacDonald of *Politics* refused to accept the expedient and glib evaluations of radioactivity: "Nobody knew how deadly or prolonged [the] radioactivity poison would be; yet they went ahead and used the bomb."[84]

iii. Fallout: Financial, Social, Moral and Bizarre

The financial community was both awed at the magnitude of the explosion and at the potential growth of increasing government control over scientific endeavors. "Some contend that government support implies restraints on freedom" decried *Business Week*.[85] It also informed its readers that hope for industrial control of atomic power exists since the "committee which seeks ways of regulating the use of atomic energy has a business man on it as an alternative to Secretary of War Stimson."[86] Neither the politics of the atom nor the value of the resource was lost to the business interests who were told that one ton of uranium equals 1,000,000 tons of coal.[87] *The New Republic* countered by saying it would be "unthinkable" for private firms or industrial interests to control atomic energy,[88] but more realized that those "with courage, imagination, and with well-organized, well-financed, well-directed research" would enjoy rich rewards.[89]

The Chesapeake and Ohio Railroad responded with half-page advertisements asking: "How many post-war jobs—from coal" and telling that coal will "continue as one of our basic sources of power."[90] But such logic could not withstand onslaughts as that which "compared the miracle of the atomic bomb with the works of God."[91] And the editors of the Washington *Post* just three days after the bombing wrote that "the financial press is already showing signs of becoming jittery over the prospect of competition between atomic power" and conventional sources of power.[92] One day earlier (August 9), the financial section of the Chicago *Tribune* noted that "Bankers will weigh more carefully the long term bonds of utilities."[93] The public was told that uranium was cheaper than TNT and that it would be used to power plants;[94] that "a few grams could power the Queen Mary from Europe to the United States";[95] that "telephone-booth sized heating plants" would provide power for great factories; and, that one could take 1000-hour auto trips on just one gram of atomic fuel.[96] The following headlines, and other similar ones, caused an optimistic groundswell for peaceful atomic power:

Prelude to Electronic Age Ushered in on July 16 Before Tense Group of Scientists at New Mexico Proving Ground.

Single Atomic Bomb Shakes Japan with Force Mightier than 20,000 Tons of TNT to Launch New Era of Peace.

President Proclaims First Use of War's Greatest Secret: Predicts Day When Gigantic Energy Will Change Industry.[97]

Energy in U-235 Termed Force to Remake Life.[98]

After all, there was cause for economic optimism. Had not the bomb's initial $2,000,000,000 investment already paid dividends by bringing about peace? Was there not a beneficial cost-profit ratio since *one* bomb saved *nine* days of conventional warfare? These statements, and others [such as "80% of the World's Uranium in U.S."][99] lead many toward economic optimism—an optimism, however, which was not shared by all.

For instance, the dean of the school of mineral industries at Pennsylvania State College, proclaimed (however inaccurately) that "the world's entire known uranium ore supply is probably too small to furnish atomic power for more than a year or two."[100] *Business Week* warned that trying to use atomic energy for peaceful purposes was "about the same as using TNT to run the family washing machine."[101] Most reservations about atomic power were echoed by General David Sarnoff: "Mankind Must Make a Fateful Decision As Atomic Weapons Appear."[102] One noted atomic scientist said "In peacetime no engineer or scientist in his right mind would consider such a step."[103] Even Harry Truman somewhat tempered his own enthusiasm by stating "Atomic energy cannot compete with coal or falling water" and by suggesting that "a longer time was needed for research."[104]

If one tired of reading "atomic news," which, incidentally, made up a markedly small percentage of the news in any newspaper or magazine, one could turn to *Life* magazine for August 13, 1945 and read about the Superbabies (the illegitimate children of the Nazi SS troops), how the Shooting Star had crashed (P-38 jet, America's first), that *Anchor's Away* was the Movie of the Week, why Byron Nelson was the best golfer alive, or how water hyacinths were clogging southern waterways ("The Hyacinth Menace"). Or if more patriotic, one could take pride in the fact that only one woman in Stuttgart, Germany, had reportedly been raped since the American occupation began.[105] The mayor of the town, however, said he didn't believe the girl's story.[106] The Americans were much more restrained than the French, however, who reportedly raped 1,000 women during the French Occupation; and they were far more civilized than the turbaned Morrocan troops who had raped over 5000 women (Senator James O. Eastland of Mississsippi was investigating this latter charge).[107]

Things were looking up: Gasoline was ample, there were tires by the millions, nylons by yuletide, girdles by January 1st, and ways for building a business on the skills of war.[108] One could read about the happy Japanese captive who, while photographed taking a shower on an American destroyer, "expressed admiration for bathing facilities and other ways of the West."[109] "38 Win Scholarships to Julliard School"—and the names were listed.[110]

One didn't have to read about the bomb. One could read how thrilled "Bud" Baker was at his first sinking of a "Jap ship."[111] Or one didn't have to read at all for *Meet Me In St. Louis* with Judy Garland, Margaret O'Brien and Mary Astor provided ample diversion. Failing that, one could see *Out of This World* with Veronica Lake. Iced watermelon sold at five cents a pound, nine ounces of French's mustard for twelve cents, sirloin steak for thirty-nine cents a pound and potatoes for seventy-five cents a peck. There was plenty to take one's mind off the atomic bombing.

One article in the Chicago *Tribune* told how "Dede De Vaine of St. Petersburg, Florida, had been named the favorite "bomb-a-dear" of the B-17 pilots at Drew Field.[112] "Sioux Indians Plan Five-Day Sun Dance in Victory Prayer"—this was only the third time in history such a dance was held. The second time was in 1942 (with no apparent results) and the first time in 1876 (just before the Battle of the Little Big Horn and the defeat of General Custer).[113] Senator Soaper says:

> Maybe we use the wrong psychology on the inscrutable Jap. Instead of dire threats, the better approach, perhaps, would be "Could I interest you in some life insurance?"[114]

And if Senator Soaper were not amusing enough, "Gracie Allen Says" would have one rolling in the aisles:

> Well, it looks as though science has gone and invented something almost as destructive as the little boy next door. It is called an atomic bomb and there is a rumor Japanese generals refer to it as the "atom's apple." Guess this is because they get a lump in their throats everytime they think about it.[115]

However, few people were laughing and fewer still joked about the atomic bombing of a major civilian center. "Astonishment and awe and a creepy feeling down the spine were Briton's first reaction," read one line of a story entitled "Future War Called 'Impossibly Terrible'."[116] One editor asked: "But can mankind grow up quickly enough to win the race between civilization and disaster?"[117] "Yet when this is said, we have sowed a whirlwind," wrote Horace W. Baldwin, "Because our bombing has been more effective and hence more devastating, Americans have become a synonym for destruction."[118] Another editor pointed out that there existed a "race between the scientific ingenuity of man and his growth as a responsible social being."[119] Dr. David Lawrence, writing in the Detroit *News*, says: "There is no sense of jubilation . . . there is a sense of horror."[120] "Either we live as one world or we shall be destroyed," warned Barnet Nover in "To What Fell Use."[121] *Newsweek* claimed the atom bomb "terrified scores of millions," "awed hundreds of millions," and caused "fear and awe in politicians and scientists who know most about it."[122] Senator Warren G. Magnuson claimed the world must either begin to live in peace or "begin the race to win World War III."[123]

But there were those who *immediately* questioned the military use of the bomb:

> Could an announcement of the tests of the atomic bomb made in New Mexico recently have been used as a dramatic means of persuading the Japanese militants to release their people and surrender?
> Surely we cannot be proud of what we have done. If we state our inner thoughts honestly, we are ashamed of it.[124]

"Nothing in history suggests that control of this weapon is possible therefore, we must look forward not back," said Bruce Bliven, writing in *The New Republic*.[125] *Business Week* responded by claiming that the atom bomb had brought an end to the war, thereby "topping-off of mounting public debt."[126] "Nothing that was done or thought of before the atomic bomb dropped on Hiroshima requires to be recast in light of this overwhelming event," wrote the editor of the Washington *Post*, and those words rang true in the ears of many who believed the same.[127] TRB's "Washington Notes" echoed the sentiments of yet another philosophical position:

> The terrible moral decision to use the atomic bomb was made by a few people. I suppose that was inevitable. Once you invest two billion dollars in a firecracker, you have to light it. Personally, I am sick and tired of decisions like this being made in secret.[128]

iv. The Atomic Bomb Through The Looking Glass

Ironic events occur daily; however, they only become so when juxtapositioned against an immense event (fig. 45-48). The atomic bombing of Hiroshima is such an event; the following are the ironic events as reported. First, Harry Truman oversleeps on the morning of August 6, 1945.

> The President was chided today by members of his party for oversleeping when he remained in bed until nearly 7:00. Press Secretary Charles G. Ross boasted it was the first time he had arrived at breakfast table in advance of the Chief Executive who usually rises at 5:30 a.m.[129]

"Bombs Away Won the $5,000 Tompkins Memorial Sweepstakes for Two-Year-Old Trotters," read one innocent sports story which appeared in the New York *Times*.[130] McGraw-Hill announced the publication of *Atomic Power*, "priced at 25¢ a copy" and "devoted to the physics of the atom."[131] The front page headline of the Detroit *News* of August 8, 1945, read as follows:

> FIX ATOM TOLL AT 100,000
> FRANKENSTEEN FIRST BY 14,000[132]

The reference is to Richard T. Frankensteen who was leading the election for Mayor of Detroit. "Atomic Bomb Termed Magnificent" was the lead to a story by a George Washington University physicist who said the "destruction [was] beyond the wildest dreams of the imagination."[133] The National Press Club Bar in Washington has left its mark on history with the following: "Today's Special: Atomic Cocktail! 60¢." (gin and absinthe).[134]

Six Japanese soldiers were having lunch, facing away from the blast, when the bomb fell on Hiroshima. Five died in the next forty-eight hours but one survived. As late as December 10, 1945, *Newsweek* carried a story that the survivor was carefully being studied by scientists—"his diet, his ancestry, his habits of life—for a clue to survival in the next war."[135]

Time listed an even more bizarre collection of responses in its November 12, 1945 issue. One scientist from the Massachusetts Institute of Technology expressed concern that an atomic weapon could be shaped like a grand piano, "when someone sat down to play it—Blam—a city is destroyed"[136] He also warned that "Science has no means of detecting atomic explosions before they are detonated."[137] One civil engineer wanted to place elevators in huge holes under large buildings to move them down

Pix Rentals From U. S. Camp Theatres Now Nearly $9,000,000 Per Annum

$30,000 Nitery Grosser

The Latin Quarter on Broadway, with its better-than-$20,000 steady business, and hitting around $30,000 over the two holiday weeks, is the sensational grosser currently of all cafes.

Lou Walters also operates a Latin Quarter in Miami in addition to his Gotham L. Q.

'A HANDSOME 4F' PICKS HIS SPOTS

'A handsome 4F' is more and more becoming something a talent scout dreams about. Result has been the skyrocketing in the past month or two of terms agents are asking film companies for any lad with a passable personality, fair to middlin' looks, a bit of acting experience—and a 4F draft card.

Motion picture industry is currently getting close to $9,000,000 a year in film rentals from the U. S. Army camp circuit of theatres, now nearing the 1,000-theatre mark, according to estimates within the trade. Army-Navy theatres get pictures first run, normal clearance protection for civilian theatres not figuring, and pictures are sold either on a low flat rate rental or percentage based on 15c admissions (12c via books of tickets). A 'Mrs. Miniver,' 'Yankee Doodle Dandy' or 'In Which We Serve' would thus be shown to service men simultaneously with and sometimes ahead of the first New York or other key engagement.

Rentals from the now important Army-Navy theatre chain (some distribs report $900,000 and more annually in billings) in large measure offset the loss in rentals in situations adversely affected by the war due to shifting population as a

(Continued on page 25)

The Thrill Is Gone, Some Of 'This Is the Army' Cast

Fig. 45
Variety, Jan. 12 1945, front page.

Wednesday, August 15, 1945 VARIETY

FOR THE RECORD:

CBS was <u>first</u> again—with news of the Japanese surrender bid

(but we're sorry we have to mention it at all)

Out of scores of "firsts" which the CBS World News staff has scored during the war, we have drawn public attention to *none*. We have purposely avoided doing so. Until now.

Today, we *reluctantly* publish the actual record of CBS leadership (by a small fraction of a minute) over all other networks in flashing to CBS listeners the news of the Japanese surrender bid.

Why reluctantly? Because CBS, along with many responsible broadcasters, knows the danger of haste in handling news. Especially transcendent news during a world-shaking conflict. There is nothing more dangerous to a war-anxious world than a hair-trigger, scoop-conscious broadcaster. Knowing that his company is averse to any public exploitation of mere speed is the best guarantee any newsman can have that integrity comes first.

Then why publish this timetable at all? Partly because CBS, which was first in fact with the Japanese surrender news, has been publicly misrepresented as second, third, or fourth.

And for two other reasons...

ACTUAL 4-NETWORK TIMETABLE

(As electrically monitored from the air waves Aug. 10)

Here is the actual split-second timing of the first news of the Japanese surrender offer, as broadcast over three of the four networks.

CBS Network: 7:36:30 a. m.
2nd Network: 7:36:45 a. m.
3rd Network: 7:39:55 a. m.
4th Network: (slightly later)

First, to make clear at a glance how superbly all networks served the American people in this final crisis of the war. The news came through at 7:35 a.m. Note that within 300 seconds it was winged over three networks at the speed of light.

Second, to let you know that CBS, in common with most broadcasters, would rather lose an hour of time than an ounce of integrity in handling news.

At the outbreak of the war, CBS re-emphasized this credo in the following instructions to its far-flung staff of newsmen at home and abroad:

"Columbia wants to be first with the news whenever it can and as often as it can, but Columbia would rather be last, or never broadcast certain material at all, than to rush on the air with news that should not have been put out."

How well this policy has served our listeners will be remembered from the furor of the false peace talk last March, when there was *no* such furor over CBS microphones. And again last April when the European war "was over"—but it wasn't.

We pledge that CBS will never let the sweep-second hand of a stop watch outweigh our grave responsibility to be accurate, first or last—but *all the time!*

THE COLUMBIA BROADCASTING SYSTEM

Fig. 46
Variety, August 15, 1945, p. 39.

Fig. 47
Newsweek, April 3, 1945, p. 45.

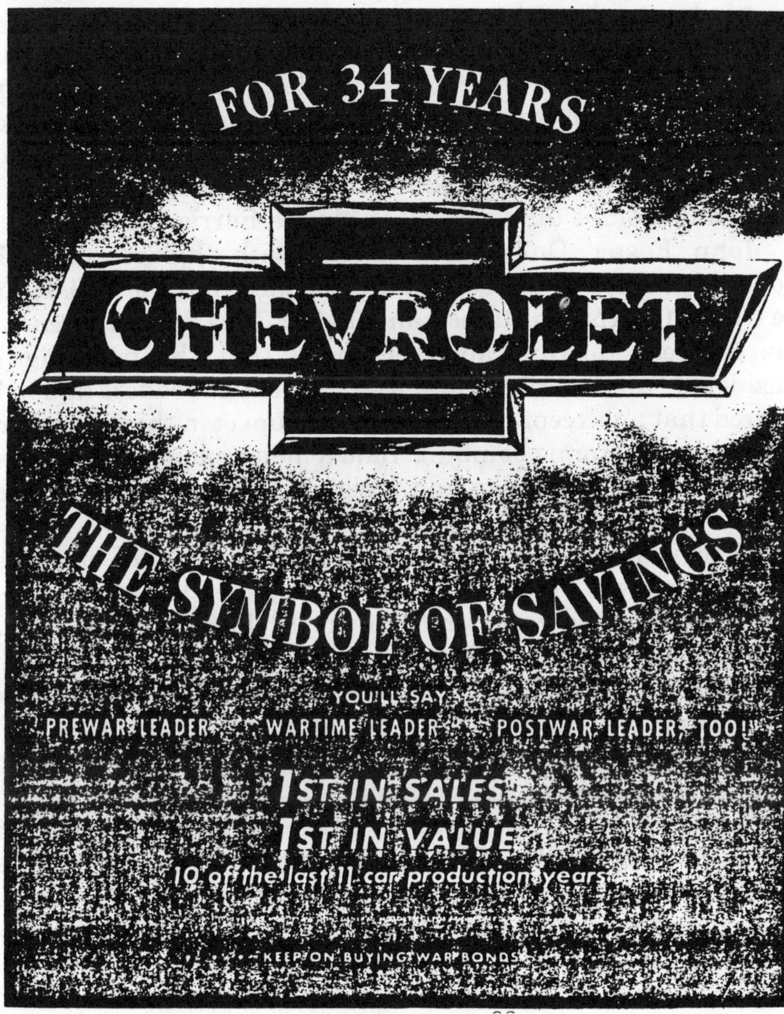

Fig. 48
Look, August 21, 1945, p. 83.

out of danger in case of atomic attack.[138] One Oak Ridge, Tennessee scientist compared atomic war to a duel, with "whoever shoots first wins."[139] Isolationist *Orphan Annie* declared that "international gangsters" had stolen Daddy Warbuck's atomic secrets. The gangsters were aided by corrupt "politicians."[140] And in *Smilin' Jack*, Saddlesoap Jones bought two atomic bombs (and a B-29) to blast a hurricane.[141]

John Foster Dulles left history with this response: he protested the use of the bomb and "pleaded that the U.S. follow the ways of Christian statesmanship."[142] *The Christian Century* was far less tolerant and demanded that the bombing be considered "an American atrocity."[143] Chicago readers had to be assured that an exceptionally dramatic sunset in the west had no relation to the atomic bombing. Said a meteorological forecaster: "Any time you get smoke and moisture in the western sky you have a red sunset."[144]

There were also times before the attack when the secret of the atomic bomb was in jeopardy, and in danger of being uncovered. In one instance, a reporter from the El Paso (Texas) *Herald* investigated reports of a huge explosion on July 16, 1945—the date of the test bombing. The official report stated that an underground ammunition dump had accidentally exploded. Such a mundane story was too bland for this enterprising reporter who went on to embellish his account with exaggerations of "the greatest fireworks show" which "illuminated whole mountain chains."[145] (Apparently, Japanese spies did not read the El Paso *Herald*—circulation 27,046). Another near disclosure occurred when reporters asked what was being built in Oak Ridge, Tennessee in October of 1944. The response was that "35,000 workers are making Roosevelt campaign buttons."[146] Even comic strips were censored to insure that the slightest hint of disclosure would not occur. In an April 14, 1945 Superman comic strip, Superman was to be bombarded by a 3,000,000 volt charge generated by a cyclotron: "Even Superman can't take it."[147] Superman had been warned.

> But Superman could take it, and did. What he couldn't take was the office of censorship which asked McClure Newspaper Syndicate to discontinue references to atomic energy. A new series of strips, then in production, was cancelled, and Superman went into a sequence in which he played a baseball game single-handed.[148]

Yet of all the ironic and, in this case, quite unexplainable coincidences of this period was the appearance of a poem which was contained in an editorial feature. The poem appeared on August 7, 1945 in the Chicago *Tribune* and is entitled "Hollyhocks":

A fireworks flower
  In my floral docket—
Shooting straight up
  Much like a rocket.

Some flowers moult
  In hues that drizzle
That swat the eye
  With colors that sizzle.

There once was a yard
  Where hollyhocks grew
So riotously
  And we did, too.

Hollyhocks still
  Are as bright today
Tho life may have lost
  A bit on the way.

But, shoot the works, hollyhocks
  Shoot a mile high
You somehow express
  The mood of July.[149]

Evan S. Tonian

August 7, 1945—most of the type for these "reader features" are set long before the front pages or the sports news. How did this poem appear at this place at this time? A literary critic could speculate on its meaning, but perhaps some rhetorical questions will suffice. Are the "fireworks" "shooting up like a rocket" a reference to the atomic bomb? What about hues and colors that sizzle? Is the yard where hollyhocks grew the University of Chicago? What of the life that may be lost on the way? "Shoot the works" and "shoot a mile high" seem like direct references to the only mile high weapon of World War II—the atom bomb. And what is "the mood of July"? Can it be an allusion to July 16, 1945, the day of the desert test bombing? And the coincidence continues with the author's name—Tonian—in that the Enola Gay which carried the bomb to Hiroshima left from the island of Tinian.

Amazing coincidence? Perhaps. And why not on a day so strange and so bizarre. The words of one editor best summarize the feeling of many who were confused on that fateful day: "At first it was a strange place, full of weird symbols and the smell of death."[150]

The Inter-American Judicial Commission found little strange or confusing about the new bomb. As early as August 10, 1945 in a meeting in Rio de Janeiro the commission declared that "the flying bomb is an unlawful weapon" (German V-1 and V-2), but the atomic bomb was lawful because it could be guided away from undefended areas.[151] It seems at least some American newspapers readers were content to be informed of this fact.

v. The Final Solution?

There was still one serious question left unsolved: Who should control the bomb and what should become of the atomic secret? Whatever group—industrial, military, political or international—believed it should control the new energy. *Newsweek* called the War Department's plan for secrecy "the most reactionary in American history"[152] and claimed secrecy would only lead to increased industrial development in Russia, since Russia would have the secret within two years.[153] The *New Republic* called for universal disclosure and international control for "every nation which does not know the secret will look at this country with fear and suspicion."[154] Besides, it argued, the United States could still get credit for turning the secret over to the Security Council.[155] Morris L. Cooke expressed concern that Congress did not fully understand the power of science and believed it was inadequately prepared to deal with "either the atomic bomb or the atomic secret."[156] *Time* took the position that the atom bomb changed nothing for "those who had the power favored keeping it, those who lacked it favored sharing it."[157]

So there they are, the reactions of the people portrayed in the press and the reactions of the press itself to the atomic event. There had never been anything like the first atomic bombing—there will probably never be anything as dramatic. It was awesome, magnificent, terrible. Some people interviewed on Christmas Eve 1945, in line for the Annual Radio City Music Hall Christmas Show, were quoted as believing the following:

> Listening to the people talk, the pollsters found awe, fear, cynicism, confusion, hope, but mostly confused hope and hopeful confusion. They found a substantial minority expectation that atomic experiments would end the world someday, a vague majority confident that somehow

everything would work out and that man would somehow be better off in the long run.[158]

However, when all is said and done, when all is printed and read, and when the jokesters and punsters and editors switch focus to a new story of a yet different technology or a more exciting personality, one reaction has burned itself forever in history—and it appeared in virtually all the newspapers which carried the story of a member of the Enola Gay's crew on the bombing of Hiroshima:

Chicago born William S. Parsons described the explosion as "tremendous and awesome" and reported that "the crew said, 'My God,' and couldn't believe what happened."[159]

Notes

[1]"Tense Truman Bares Secret," Chicago *Tribune*, August 7, 1945, p. 1.
[2]The New York *Times*, August 1, 1945.
[3]*Ibid.*
[4]New York *Times*, August 1, 1945, p. 13.
[5]Washington *Post*, July 1, 1945, p. 5.
[6]Washington *Post*, August 4, 1945, p. 1.
[7]Washington *Post*, July 30, 1945, p. 1.
[8]Washington *Post*, July 31, 1945, p. 1.
[9]New York *Times*, August 3, 1945, Vol. XCIV, No. 31, 967.
[10]New York *Times*, August 2, 1945, p. 27, Financial Section.
[11]Washington *Post*, July 29, 1945, Section II, p. 1B.
[12]New York *Times*, August 2, 1945, p. 1, Vol. SCIV, No. 31, 967.
[13]New York *Times*, August 2, 1945.
[14]New York *Times*, August 3, 1945, p. 13.
[15]Chicago *Tribune*, "Sentry Fires on Jack Benny, 3 More in Car," August 1, 1945, p. 5.
[16]Chicago *Tribune*, August 1, 1945, p. 1.
[17]New York *Times*, August 2, 2945, "Desperate Famine Facing Vienna, Transport Gone, Food Unavailable," p. 12.
[18]Washington *Post*, August 4, 1945, "Yanks Burned Alive by Japs, Letter Reveals," p. 2.
[19]Chicago *Tribune*, August 4, 1945, p. 9.
[20]Chicago *Tribune*, August 4, 1945, Entertainment Section.
[21]New York *Times*, August 4, 1945.
[22]Chicago *Tribune*, August 1, 1945, p. 12.
[23]Chicago *Tribune*, August 2, 1945, p. 16.
[24]New York *Times*, August 3, 1945, Financial Section.
[25]Washington *Post*, August 4, 1945, p. 1.
[26]Washington *Post*, August 1, 1945, p. 2.
[27]New York *Times*, August 4, 1945, Vol. CIV, No. 31, 969, p. 1.
[28]Washington *Post*, August 1, 1945, Editorial section, p. 8.
[29]New York *Times*, August 2, 1945, p. 36.
[30]Detroit *News*, August 6, 1945, p. 1.
[31]Ann Arbor *News*, III, No. 187, August 6, 1945, p. 1 (headline).
[32]Detroit *News*, August 6, 1945, p. 1 (headline).
[33]Ann Arbor *News*, III, No. 187, p. 5.
[34]Chicago *Tribune*, August 7, 1945, p. 1 (headline).
[35]Ann Arbor News, August 7, 1945, "Battle of Laboratories Won by Allied Scientists," p. 4.
[36]Detroit *News* August 6, 1945, p. 1.

[37]Chicago *Tribune* August 7, 1945, p. 1.

[38]*Ibid.* [39]*Ibid., p. 8.*

[40]Hightower, John M., "Tokyo May Be Next Target for Explosives," Ann Arbor *News*, August 7, 1945, p. 5.

[41]New York *Times*, August 7, 1945, p. 1.

[42]*Ibid.*, p. 7.

[43]Chicago *Tribune*, August 7, 1945, "Hiroshima, No. 1 Target, Turns Out Guns, Tanks," p. 4.

[44]Chicago *Tribune*, "Atomic Bomb Crew's Story," p. 1.

[45]Ann Arbor *News*, August 8, 1945, editorial page.

[46]Washington *Post*, August 8, 1945, "Richenbacher Sees Air Power Beating Japan," p. 16.

[47]*Newsweek*, XXVI, No. 7, August 13, 1945, pp. 30-32.

[48]"Future of the Atomic Bomb," *U.S. News*, August 17, 1945, XIX, No. 7, pp. 20-33.

[49]*Life*, Vol. 19, No. 8, August 20, 1945, p. 25.

[50]Ann Arbor *News*, August 6, 1945, p. 1.

[51]"Heard Around the World," New York *Times*, August 7, 1945, p. 22, Editorial Section.

[52]*U.S. News, Ibid.*

[53]*Ibid.*

[54]"Ann Arbor *News, Ibid.*

[55]"Atomic Bomb Statement is Broadcast by Moscow," New York *Times*, August 7, 1945.

[56]"Texts of the Day's Communique on Fighting in Various Zones, *Japan,"* New York *Times*, August 8, 1945, p. 2.

[57]Detroit *News*, August 7, 1945, p. 1.

[58]"Churchill Hails Roosevelt in Discovery," Washington *Post* (London), August 6, 1945, p. 1.

[59]New York *Times*, August 9, 1945, p. 11.

[60]"Benevolent Japs Won't Use 'Atom Weapon' on U.S. Civilians," Washington *Post*, August 10, 1945, p. 3.

[61]*Ibid.* [62]*Ibid.*

[63]"Paper Hits Use of Atomic Bomb," Washington *Post*, August 10, 1945, p. 5.

[64]*Ibid.*

[65]"Turks Cite Atomic Bomb," New York *Times*, August 13, 1945, p. 8.

[66]*Time*, August 20, 1945, p. 36.

[67]*Ibid.*, p. 42.

[68]"Curtain Rises on War's Last Act as MacArthur Takes Over Japan: Atomic Agony," *Newsweek*, Sept. 3, 1945, pp. 19-28.

[69]New York *Times*, August 7, 1945, p. 5.

[70]"5000 Scientists at U of C Worked Three Years on Bomb," Chicago *Tribune*, August 7, 1945, p. 1.

[71]*Ibid.*, p. 8, editorial.

[72]Chicago *Tribune*, August 8, 1945, p. 4.

[73]Lindley, E.K., "Where Credit is Due," *Newsweek*, August 20, 1945, p. 56.

[74]Washington *Post*, August 9, 1945, p. 1.

[75]*Ibid.*

[76]"The Defeated," *Time*, Sept. 3, 1945, p. 28.

[77]"Black Lesson of the Atom Bomb: It's Awful Fury Even Kills Hope," *Newsweek*, XXVI, No. 25, Dec. 17, 1945, p. 37.

[78]"Bomb's Destructive Effects a Most Scientific Question," Detroit *News*, August 8, 1945, p. 1.

[79]"Scoff at Fear of Atomic Blight," Chicago *Tribune*, August 9, 1945, p. 6.

[80]"Atomic Energy: No Cause for Worry," Detroit *News*, August 8, 1945, p. 14.

[81]"Scientist Hits Atomic Superstition," Washington *Post*, August 9, 1945, p. 6.

[82]"The Defeated," *Time*, Sept. 3, 1945, p. 28.

[83]*Ibid.* [84]*Ibid.*

[85]"Washington Bulletin: Science at a Crossroads," *Business Week*, No. 832, August 11, 1945, p. 5.

[86]*Ibid.* [87]*Ibid.*

[88]Bliven, Bruce, "The Bomb and the Future," *The New Republic, Vol. 113, No. 8, August 20, 1945, pp. 37-38.*

[90]"Policies and Principles: First Atomic Returns," *Time*, August 27, 1945, p. 39.

[91]"Jesuit Speaks on Atomic Bomb," Washington *Post*, August 10, 1945, p. 19.

[92]"Atomic Power," Washington *Post*, August 10, 1945, p. 14, editorial section.
[93]"Atomic Force Appraised for Trade Effect," Chicago *Tribune*, August 9, 1945, p. 27, financial section.
[94]Chicago *Tribune*, August 8, 1945, p. 4.
[95]"Terrific Power: Scientists See Start of Atom Era," Washington *Post*, August 7, 1945, p. 3.
[96]*Ibid.*
[97]Washington *Post*, August 7, 1945, p. 1.
[98]Chicago *Tribune*, August 7, 1945, p. 2.
[99]Washington *Post*, August 8, 1945, p. 2.
[100]"Atomic Power Limited by Ore Shortage," Chicago *Tribune*, August 11, 1945, p. 7.
[101]"Washington Bulletin: Science at a Crossroads," *Business Week*, August 11, 1945, p. 15.
[102]Sarnoff, David, "Science for Life or Death," New York *Times*, August 10, 1945, p. 6.
[103]Jaffee, Bernard, "How the Bomb Came to Be," *New Republic*, Sept. 17, 1945, p. 344.
[104]"The Nation: Birth of an Era," *Time*, XLIV, No. 7, August 13, 1945, p. 17.
[105]"Rape Story Dispute Grows in Stuttgart," New York *Times*, August 11, 1945.
[106]*Ibid.* [107]*Ibid.*
[108]*Business Week*, August 18, 1945, p. 22.
[109]"Jap Discovers Western Luxury," Washington *Post*, August 9, 1945, p. 3.
[110]New York *Times*, August 7, 1945.
[111]"Sinking First 'Jap' Ship Thrill for 'Bud' Baker," Ann Arbor *News*, August 7, 1945, p. 5.
[112]"Bomb-a-Dear," Chicago *Tribune*, August 6, 1945, p. 7.
[113]*Ibid.*, p. 12.
[114]"Senator Soaper Says," Detroit *News*, August 8, 1945, p. 20.
[115]"Gracie Allen Says," Detroit *News*, August 9, 1945, p. 30.
[116]"Atom Items," Detroit *News*, August 7, 1945, p. 1.
[117]New York *Times*, August 7, 1945, p. 22, editorial section.
[118]Baldwin, H.W., "The Atomic Weapon," New York *Times*, August 7, 1945, p. 10.
[119]"We Must Prevent Another War: Man Can Now Destroy the World," Detroit *News*, August 7, 1945, p. 10, editorial page.
[120]Lawrence, David, "Sense of Horror Attends Success of Atomic Bomb," Detroit *News*, August 8, 1945, p. 20.
[121]Nover, Barnet, "To What Fell Use," Washington *Post*, Augsut 9, 1945, p. 8.
[122]"The Periscope: The Power of the Universe," *Newsweek*, XXVI, No. 8, August 20, 1945, p. 34.
[123]*Ibid.*, "A New Era: The Secrets of Science."
[124]"Lawrence, David, "What Hath Man Wrought?" *U.S. News*, XIX, No. 7, August 17, 1945, pp. 211-213.
[125]Bliven, Bruce, "The Bomb and the Future," *The New Republic*, Vol. 113, No. 8, August 20, 1945, pp. 211-213.
[126]"Lighting Up the V-J Question," *Business Week*, August 11, 1945, p. 120.
[127]"Second Thoughts," Washington *Post*, August 9, 1947, p. 8.
[128]"TRB Washington Notes," *The New Republic*, Vol. 113, No. 8, August 20, 1945, p. 222.
[129]"Truman Sings Simple Hymn with Sailors," Washington *Post*, August 6, 1945, p. 4.
[130]New York *Times*, August 10, 1945, sports section.
[131]"Has Its Own Journal," *Business Week*, No. 833, August 18, 1945, p. 22.
[132]Detroit *News*, August 8, 1945, p. 1.
[133]"Atomic Bomb Termed Magnificent," Washington Post, August 8, 1945, p. 2.
[134]Washington *Post*, August 8, 1945, p. 3.
[135]"The Atomic Man," *Newsweek*, Dec. 10, 1945, XXVI, No. 4, p. 65.
[136]*Time*, VLVI, No. 20, Nov. 12, 1945, p. 29.
[137]Ibid. [138]*Ibid.* [139]*Ibid.* [140]*Ibid.* [141]*Ibid.*
[142]*Time*, XLVI, No. 16, Oct. 15, 1945, p. 62.
[143]*Ibid.*
[144]"Two Cloud Layers In Storm Cause Unusual Sunset," Chicago *Tribune*, August 5, 1945.
[145]"The Near Misses: The Big Bang," *Newsweek*, August 20, 1945, p. 68.
[146]*Ibid.*, "The Button Factory."
[147]*Ibid.*, "The Superman Way."
[148]*Ibid.*
[149]"A Line O'Type on Two: Hollyhocks," Chicago *Tribune*, August 7, 1945, p. 8.

[150]*Time*, August 20, 1945, plate 29.

[151]"Atomic Bomb is Lawful Jurists Decide at Rio," New York *Times*, August 11, 1945, p. 5.

[152]"The Periscope, *Newsweek*, XXVI, No. 13, Sept. 24, 1945, p. 25.

[153]*Ibid.*

[154]"Atomic Bomb Secrecy," *The New Republic*, Vol. 115, No. 12.

[155]*Ibid.*

[156]Cooke, Morris L., "Congress Grapples with Science," *The New Republic,* Vol. 113, No. 15, Oct. 1, 1945, pp. 433-435.

[157]"Policies and Principles: First Atomic Returns," *Time*, August 27, 1945, p. 30.

[158]"National Affairs: The People," *Time*, XLVI, No. 27, Dec. 24, 1945.

[159]"Chicagoan Tells of American Scene from Plane," Chicago *Tribune*, LIV, No. 119, August 8, 1945, front page.

Chapter IV
The Salk Vaccine: "The Biggest Experiment in U.S. Medical History"

Life, *1954*

Early in 1953, readers of the *Science News Letter* were informed in a somewhat objective and understated manner that a vaccine developed by Dr. Jonas Salk had been judged effective in its initial experimentation against poliomyelitis (fig. 50). However, Dr. Salk carefully defused any premature suggestion that polio could be quickly eliminated by indicating that such results should not be taken "to mean that a practical vaccine was now at hand."[1] By autumn of 1954 the attitude toward the new vaccine had become less passive and the early signs of justifiable exuberance—still premature—became manifest in statements by the media which promised an effective vaccine within a year.[2] *Newsweek* optimistically declared that "enough polio vaccine should be available commercially in 1955 to immunize all children in the United States."[3] Somewhat balancing were more temperate reports that the vaccine was still experimental; and, despite comments of no unfavorable results from six hundred adults who were tested and that the immunity lasts for several months, "routine immunization for polio" would not be "available for some time."[4]

No reports, however, could in fact temper the expectation of a majority of American readers who believed that soon, and perhaps very soon, a victory against the crippling effects of polio would be achieved. In essence, the battle was engaged and the public expected that humans armed with medical science and advanced technology would easily and quickly defeat the menacing virus. Indeed, victory did occur, but in no manner was it as simple or as rapid as some anticipated.

i. *"For Investigational Use Only,"*[5] Walter Winchell and Strange Goings-On for 1954

The significant question surrounding the vaccine in 1954 was whether it was safe enough to use on a mass scale. For most of the public there was no clear answer and proclaimed experts tended to confuse rather than clarify the issue (fig. 50). Early in 1954, Dr. Thomas Francis, Jr., from the University of Michigan School of Public Health, was reported to have said:

SCIENCE NEWS LETTER for April 4, 1953

MEDICINE

Polio Vaccine Effective

Tests of three-way polio vaccine on 161 persons give "encouraging" results. Its production will be delayed until safety trials have been completed.

➤ A VACCINE effective against all three types of poliomyelitis has come through trials on 161 persons with "encouraging" results.

The vaccine brought the anti-polio substances in the blood to as high a level as that induced by an attack of paralytic polio. In some cases a higher protective level was reached.

The vaccine took effect within six weks. The anti-polio substances, or antibodies, have remained at the high level for four and a half months. This is the longest period studied but presumably the protection will last even longer.

These results are announced by Dr. Jonas E. Salk of the University of Pittsburgh School of Medicine in the *Journal of the American Medical Association* (March 28).

The results should not, Dr. Salk states, be taken to mean that a practical vaccine is now at hand.

(The National Foundation for Infantile Paralysis in New York, which aided Dr. Salk's studies, says no plans have been made yet for field trials of the vaccine.)

The chief reason for the delay in practical vaccine production is the safety factor. No signs of illness that could be attributed to the vaccine have occurred in any of the persons who got it. A few had some redness and swelling for 24 to 48 hours at the site of the vaccination.

But great care and "considerable time" are needed to prepare and study each new

Another vaccine was made from formaldehyde treated virus incorporated in a water-in-oil emulsion and was injected into the muscles. This preparation apparently gives better results. The oil emulsion, as has been found in influenza vaccine production, increases the anti-body, or protective, response.

The tests of the vaccine were made on patients at the D. T. Watson Home for Crippled Children, Leetsdale, Pa., and at the Polk State School, Polk, Pa. Some of those given the vaccines were patients who had been paralyzed by polio in recent years. They had, as a result, immunity to the virus type which paralyzed them but not to other types.

Collaborating with Dr. Salk in the studies were: Maj. Byron L. Bennett, U.S.A. (ret.), Dr. L. James Lewis, Miss Elsie N. Ward and Dr. J. S. Youngner, all of the University of Pittsburgh School of Medicine.

Science News Letter, April 4, 1953

Fig. 49
Science Newsletter, April 4, 1953.

It is not known at present whether the vaccine to be used is highly effective, moderately effective or ineffective in protection of human subjects.[6]

Such statements caused some concerned parents to balk at the "human guinea pig" implications because of the experimentation obvious in the earliest innoculation programs. And little solace was found in reports that "some health and medical authorities continued to unofficially criticize the entire vaccine trial program."[7] Most did so because they believed there were not enough scientific controls. In addition, there were delays in the implementation of the program because of "production delays which apparently could not be solved."[8] Confusion was proliferated by "good news—bad news" press stories which indicated that the commercial manufacturers planned no profits from the upcoming polio vaccine season. In essence, the "no profit" scheme was restricted to the early programs, which in effect were not trial periods for innoculated human beings but trial-and-error periods for the manufacturers themselves in terms of refining their production capabilities.

Further confusion was caused by an announcement from the president of the American Medical Association who suggested, in gross overstatement, that "the whole medical profession is anticipating the immediate end of polio."[9] And those who insisted on more logical and accurate evaluations of the vaccine's effects were not to be satisfied, for in reality no one yet had expressed a clear and accurate answer on either the effectiveness or the safety of the vaccine (fig. 51). Circumlocutions seemed the order of the day. Witness one interview with the Director of the National Foundation for Infantile Paralysis:

Q: Is the new vaccine safe?
A. The matter of safety is certainly one problem that I think should have been answered to everyone's satisfaction. Of course, there may be a few who will always raise questions, but we have to ask ourselves: What can be safer than safe?
Q. Is there any risk from a tame virus getting wild again or from a killed virus not actually being dead?
A. I think you have to go back to how dead is dead.[10]

Such bureaucratic mumblings by "experts" did nothing to belay the anxious public whose anxiety seemed steadily increasing. One public figure almost created mass hysteria with this announcement:

avoiding ans.

Winter Shots Avert Polio

Babies and children taking diphtheria and whooping cough shots during the summer and early fall are more susceptible to polio.

➤ TO avoid danger of infantile paralysis, babies and children should not be given "shots" against diphtheria and whooping cough during summer and early fall, when infantile paralysis is likely to appear.

This warning was sounded at the sixth International Congress of Pediatrics in Zurich, Switzerland. It was based on findings of a number of doctors in England and Australia. The baby and child specialists who devoted a special section to this subject considered it one of the most significant of the Congress.

Babies routinely immunized against diphtheria and whooping cough contracted infantile paralysis more often than other children, Drs. Philip R. Evans and J. Kenneth Martin of Guy's Hospital, London, and Dr. B. P. McCloskey of Adelaide, Australia, discovered independently in the summer of 1948.

The paralysis, these doctors found, nearly always affected the arm or leg where the "shots" had been given.

The presumption is that had it not been for these "shots" the poliomyelitis virus might have produced either no visible disease or else non-paralytic disease. The injections, however, lowered the resistance of the child locally, that is, in the arm or leg, or else set up conditions favorable to the local multiplication or spread of the virus, resulting in local paralysis.

Reports on the link between polio and whooping cough and diphtheria immunizations were given by Dr. Evans and Dr. W. H. Geffen of St. Pancras Hospital, London.

Fig. 50

Science Newsletter, August 5, 1950.

Attention all doctors and families in the United States .... The Public Health Service tested ten batches of this new polio vaccine. The government's health department, I am told, found that seven of the ten contained live not dead polio virus. It killed several monkeys.[11]

This information supplied by Walter Winchell caused massive public concern over the vaccine's safety. People asked if the report was indeed accurate. If so, why did a television commentator and not the government make the information public? After all, was not the government supporting the trial innoculations? Questions without answers, confusion heaped upon confusion, and reactions from the vaccine's supporters characterized the news of the next few weeks.

Dr. Salk responded that the only satisfactory vaccine was that which contained only dead virus and that only vaccine which has been passed "in three laboratories have been used in innoculations."[12] In stronger terms, Salk implied that such "sidewalk superintendents" were engaging in "a misrepresentation of what we are doing."[13] Reaction against the Winchell report intensified. *Time* called him "gossipist Walter Winchell" and then suggested that rejection of trial batches demonstrated that "the tests are doing what they were set up to do!"[14] Parents were warned that they may be defeating their own desires by "raising too much excitement about the vaccine safety question."[15] The general consensus among vaccine supporters was that "no state has pulled out because of Winchell."[16] However, at least one county in Michigan, Washtenaw, had decided to withdraw from the study because of the "ill-advised television broadcast by Walter Winchell on Sunday night."[17] Washtenaw—the very county in which the scientists at the University of Michigan were to rule on the vaccine's effectiveness withdrew! In a more explicit statement the Medical Society in Michigan stated: "So far we have not had an iron clad assurance that the vaccine is safe."[18] Another area included in the study, a parrish in Louisiana, also decided to delay the polio tests based upon the Winchell remark and upon the lack of assurance from health authorities that the "serum" [sic] was safe.

As the week following the television announcement progressed, readers learned of other areas which, for a variety of stated reasons, were not participating. Massachusetts, Minnesota, Maryland and North Dakota were not participating in the study. The City Medical Society of Alexandria, Virginia,

U. S. News & World Report

NEW POLIO VACCINE: YOUR QUESTIONS ANSWERED

Here, from men who know, are answers to practical questions on the new polio "shots":

- **Even one shot will give some immunity.**
- **There's no danger that the vaccine will cause polio.**
- **Immunity might last indefinitely.**
- **Children vaccinated in the 1954 tests need "boosters."**
- **Shots this year will be far more effective.**
- **Vaccine seems to prevent all polio deaths.**
- **Shots can be given to babies at 3 to 6 months of age.**
- **Adults can take shots when available.**
- **Injections do not hurt, have a water base.**
- **Immunity starts 6 to 9 days after a shot.**
- **Those who have already had polio need shots, too.**
- **Shots may be 100 per cent effective soon.**

Just how effective is the new polio vaccine?

Very effective against the most crippling types of polio. Not quite as effective against the less crippling types. Not effective against types that do not cripple.

Dr. Thomas Francis, Jr., who tabulated the results of the 1954 vaccine field tests, concludes that an individual's chance of getting paralytic polio, if vaccinated, is only 10 to 20 per cent as great as his chance would be without "shots."

Dr. Jonas E. Salk, who developed the vaccine, says that the "nonparalytic" type of polio does no lasting damage, is not considered a major problem. Nearly everyone has one or more cases during a lifetime, usually without realizing it.

Will this year's shots be more effective than those of last year's test?

Yes, for two reasons: Dr. Salk has learned a better way to prepare the vaccine and a better way to administer it. Shots given this year, in other words, will be far more potent than some given in the mass experiment last year. And the timing and method of giving shots will also insure a bigger build-up of antibodies which combat polio in the body.

What improvement is there in the vaccine itself?

Last year, the vaccine contained a chemical that was added to preserve it during storage. It was learned later that this chemical reduced the potency of the vaccine, resulting in less polio immunity. Now the vaccine is being manufactured with a better preservative or with no preservative at all.

What is the new method for administering vaccine?

Dr. Salk recommends that two inoculations be given within two to four weeks of each other, with a third shot about seven months later. Last year, all three shots were given within a five-week period. It was found that inoculations given in this short period provided "primary immunization." Only a limited number of antibodies were produced. But an additional shot after a seven-month wait made the antibodies multiply very rapidly.

How long will this immunity last?

For "an indefinite period," perhaps

Fig. 51

MEDICINE

Mass Polio Test

During the week of Feb. 8, 1954, the first public trials of the new antipolio vaccine, developed by Dr. Jonas E. Salk of the University of Pittsburgh, will get under way in one or more Southern states. This mass inoculation program, sponsored by the National Foundation for Infantile Paralysis, in which some 700,000 to 1,000,000 second-grade school children in 200 different localities in the United States will receive polio-prevention shots, will be continued to June, when the case rate for this disease starts to rise.

While all the specific inoculation centers have not yet been selected, plans have been made by the foundation for entire communities to take part in the campaign to prove the value of the new vaccine. In each area, polio authorities will be aided by public-school systems, local medical societies, health officers, and practicing pediatricians.

In the tests, each second-grade child will get three shots (one c.c. each in a water solution), the first two, one week apart; the third, four weeks later. Non-injected youngsters of the first and third grades will be used as control subjects.

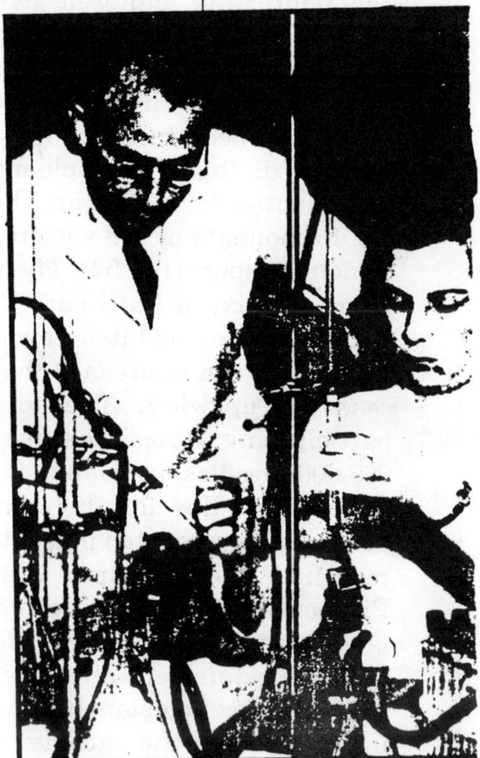

International

Salk's magic bullet hits polio

Fig. 52

Newsweek, Nov. 23, 1953, p. 62.

said that the city would not take part in the test. Was that where the politicians' children lived, some readers wondered. The *Detroit Free Press* expressed irritation at the suggestion that the vaccine be discussed in medical journals and not in the press, for "the children's parents have a right to know. Walter Winchell may have been fully wrong, as he sometimes is," the editorial continued, "but the people must be adequately informed unless the scientists' children are the only ones to get the vaccine."[19]

Proponents of the vaccine tried to ignore the reaction to the Winchell report (fig. 52). The nation was informed that Randall Kerr, age six, of MacLean, Virginia, was the first child in the nation to be innoculated. It was also revealed that no vaccination would be given in areas of the country where polio had begun its seasonal upswing. The total cost of the program in the participating states was quoted at $7,500,000 and there were 269,000 needles available for use.[20] The Polio Pioneers had permission slips for their parents to sign; of the total, sixty percent consented, ten percent refused, and thirty percent did not respond.[21] The children who did participate received "Polio Pioneer" buttons; and, indeed, there were an ample number of photographs in the press to abundantly cover V-day (vaccination day). One statement attempted to draw the Winchell comment to a definitive conclusion; however, it did little to inspire confidence in the safety of the vaccine:

> After carefully checking the reaction of children thus innoculated, the committee (Vaccine Advisory Committee) put an end to the question, raised notably by Walter Winchell, that live virus might have been found in some batches of the commercially produced vaccine. This had caused anxiety and confusion among parents, teachers and even a few health authorities.[22]

So apparently things had gotten back on schedule and normality ruled. And why not? There were plenty of other occurrences and events which had to be confronted. In addition to a best-selling book entitled *How to Lie with Statistics*—a book supposedly designed for businessmen in order to inform them on how not to be deceived by criminals[23]—one could mull over the logistics in which "52 Tons of Roquefort Cheese Made 5 Ocean Crossings."[24] Martin Block was on WABC with the "Make Believe Ballroom," or one could go to the movies and view any or all of the following: *Heidi, Living Desert* (Disney), *How to Marry a Millionaire* (Marilyn Monroe), *Julius Caesar* with James Mason

SCIENCE NEWS LETTER *for October 24, 1953*

The "Kinsey Report" on women...

SEXUAL BEHAVIOR IN THE HUMAN FEMALE

By Alfred C. Kinsey, Wardell B. Pomeroy, Clyde E. Martin, Paul H. Gebhard, and others on the staff of the Institute for Sex Research at Indiana University.

This is the "Kinsey Report" on women—the interesting, very readable study of female sexual behavior, of the biologic and sociologic factors which influence it, of the ways in which it is similar to or different from male behavior, and of its social and legal implications.

COMMENTS

"... a brilliant and arguable contribution for which we are all in debt."

—New York Times

"... can do much to make husbands and wives better understand each other."

—Collier's

"(Kinsey's message) is perhaps the most important discovery ever made in the field of human behavior."

—Life

842 Pages, $8.00 at your bookseller

Fig. 53
Science Newsletter, Oct. 24, 1953, p. 267.

and Marlon Brando, Audrey Hepburn's *Roman Holiday,* Hitchcock's *39 Steps* or *King of the Kyber Rifles* with Tyrone Power and Terry Moore (fig. 53). Isaac Stern played Beethoven and received this review:

It [the audience] cheered Mr. Stern. It cheered Mr. Szell and the Philharmonic at the end. And, undoubtedly, it was cheering Beethoven, too. Strange goings on for 1954, but such is human nature.[25]

Strange? Perhaps—but no stranger than 398,000 Mobilized Guard Units having a surprise alert, or a modern revision of Aristophanes' *Lysistrate*, entitled *Lisa Stratter* at the Provincetown Playhouse, or a proposal by Representative Thurmond Chatham that the United States swap cows for Soviet sables as a gesture toward international peace.[26] And how strange indeed must have sounded the words of Robert Kennedy (D-Mass) when he warned that "to pour money, material, and men into the jungles of Indo-China without a remote prospect of victory would be dangerous, futile, and self-destructive."[27]

President Eisenhower proposed that eighteen-year-olds have the right to vote. Ox cars were being used in Formosa to service F-84 Thunderjets. The British Broadcasting Company was being sued for showing the Mass on television, with the plaintiffs claiming it was illegal since the Mass had been abolished by the Reformation. The Vatican did not seem particulary disturbed since it had already established that seeing Mass on television did not fulfill the Sunday church attendance requirements. Chicken sold for nineteen cents a pound, pot roast cost twenty-nine and butter sold for forty-nine. Doris Day starred in *Calamity Jane*, John Wayne as *Hondo* and Kathyrn Grayson in *Kiss Me Kate* (with Howard Keel). Probably the best offer was *From Here to Eternity* with an extra Mr. Magoo cartoon.

"Tracking the Killer" is not a screen play of 1954; it is, in fact the title of an article which appeared in *Life* on August 13, 1954. Written by Robert Coughlan, it gave the reader a dramatic account of the history of the attempts to defeat polio. Coughlan calls the battle a "great medical mystery-drama, with a cast of thousands," the biggest experiment in U.S. medical history," and one of the "great medical detective stories of our time." In addition to the dramatic, the article contains information on the 1936-1937 Atlantic and Toronto experiments where 10,000 children received nasal sprays which were to prevent polio—

unfortunately "the only results being that they lost their sense of smell." More tragic was the 1935 experiment where 17,000 children were given diluted polio virus from the ground spinal cords of monkeys. The virus was mixed with a derivative of castor oil. The experiment failed miserably with at least twelve children getting polio and at least six dying; no one will ever know for certain since there was no effective monitoring of the experiment or its results.[28]

Late in 1954 with the vaccine's safety still not insured, such dramatic writing did indeed *inform* but did little to soothe. According to "Polio Gamble," The National Foundation for Infantile Paralysis was ordering 25,000,000 cc at a $9,000,000 cost for distribution.[29] In light of the failures of the past and the uncertainty of the future, people were still anxious and concerned. The NFIP expected to vaccinate more than fifty million "young Americans under the age of 18."[30] Polio virus, which had been termed "an especially bad actor" by three United States scientists who received the Nobel Prize, still had not been fully controlled by the end of 1954.[31] It is curious indeed that what the next year brought forth both fulfilled the greatest expectations of Salk and the unholy prophecy of Winchell.

ii. 1955: Great Expectations

Despite international and domestic tensions, the American public had at least one expectation that it hoped would be fulfilled in a positive sense: medical technology's wonder vaccine would prove effective against polio. Perhaps a review of the opening at the Ballet Theatre best captured optimism; of Alicia Alonzo's performance, one critic claimed, "It was one of those evenings when everybody shouted and bravoed at the slightest provocation, there were carloads of flowers."[32] In addition, economic gains had been predicted for many industries during the coming year, especially so for the movie industry. Such news usually was accompanied by emotion-rousing statements, like the following by the president of the Motion Picture Association: "There is no more justification for the governments of free men to bridle the screen than there is to curb the press."[33] President Eisenhower, in presenting a citation to Jonas Salk, spoke what many Americans believed: "There is no one in the world, from Mid-Manhattan to Mid-Siberia, who may not be a little happier because of what Dr. Salk has done."[34]

The only problem left to manage was that of production and

distribution of the vaccine since the "American people joyously accepted the good news" that the vaccine was safe and effective. The weary waiting was over, and the optimistic drug manufacturers had "gambled millions by plunging into full production."[35] The seventeen years of research and the $22,400,000 spent were now to pay dividends.[36] Even without official confirmation of the vaccine's effectiveness, the companies were "manufacturing the vaccine by the tankload."[37] Although the companies would not reveal how much vaccine had been prepared, some elements in the press stressed that killing the polio virus turned "the deadly brew into a safe vaccine."[38]

"Science has enriched mankind with one of its finest gifts" represents a typical statement found in various media.[39] "Again science has come forward to save man from a dread disease" claimed the voice which speaks for the consummate technocrat, a science which is "not the hand of man but a divine Providence which guides us toward new accomplishments."[40] Such accomplishments seem predetermined by the same hand that brought victories over yellow fever, tuberculosis and the like. In fact, there was the possibility of polio protection for life with only "minor" side-effects expected. Vaccine supporters were bolstered by the news that the French planned to manufacture and retain the vaccine for emergencies and orders were received from Sweden, Switzerland, the Netherlands, Denmark and Canada. Mrs. Roosevelt cautioned March of Dimes' supporters not to abandon the organization for there were still people who needed medical care. She made this plea after putting flowers on her husband's grave on the 10th anniversary of his death, which had been hastened by polio.[41]

The question-answer format, still popular with the media, provided absolute answers for a theretofore bewildered public: the vaccine was proclaimed "effective and incredibly safe." The answer to the question as to whether there was any danger of catching polio from the vaccine was "Absolutely none."[42]

Of course, part of the responsibility of the assumed success was due to the media's own reporting which some maintained had written medical history in a Hollywood atmosphere.[43] *Business Week*, in characteristic understatement, reported that the vaccine "is probably better known for its formal debut than any other vaccine in history" and that the "plain noisiness of the final report got news coverage almost rivaling a presidential election."[44] The vaccine was touted as evidence of the success of American "team play," a "touchdown" that resulted from "a long

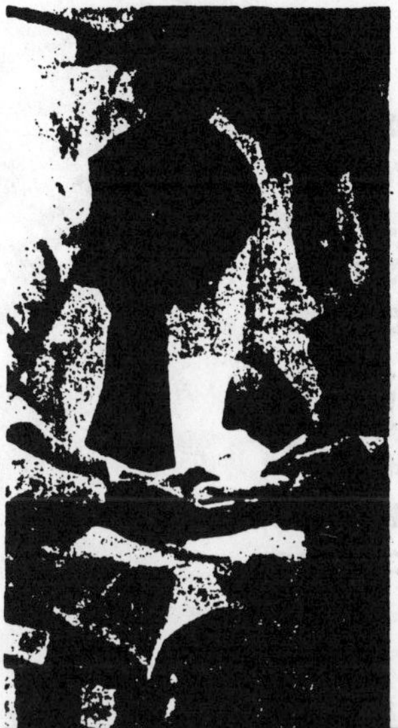

HOW THE SALK VACCINE MEASURED UP

Number of children in the test. . . .	**1,829,916**
Getting the Salk vaccine	441,131
Getting "dummy" shots	201,229
Getting no shots at all	1,187,556

Polio cases among children in the test

For those getting the vaccine	**3 in 10,000**
For those getting "dummy" shots	**7 in 10,000**
For those getting no shots at all	**5 in 10,000**

VACCINE'S PROTECTION was most effective — 94 per cent — in the type of polio most often fatal . . . ALSO, among children getting polio, lasting paralysis was less frequent for those who had received vaccine

36 U. S. NEWS & WORLD REPORT, April 22, 1955

Fig. 54

U.S. News and World Report, April 22, 1955, p. 36.

II. Cherry Pop

The next phase is production. In principle, production of Salk's vaccine is simple (pictures, pages 136-7). The virus is grown on cultures of monkey kidney tissue. It is then harvested, inactivated, and packaged. In practice, the process is complicated—partly because of the difficulties of tissue culture outside the living organism, and partly because of the newness of the technique.

• **Many Monkeys**—It starts with monkeys—lots of them. One estimate is that we will need 100,000 annually.

Rhesus monkeys from India are used most frequently for actual production. Cynomolgus monkeys from the Philippine Islands are used chiefly for testing. In the last week before the results were announced, a crisis developed in the monkey market. On a stopover in England, 390 monkeys were accidentally suffocated. India placed an immediate ban on exports of Rhesus monkeys. Representatives of the National Foundation, Eli Lilly, and Parke Davis are now in New Delhi trying to get the ban lifted.

Monkeys are essential for three reasons: (1) Viruses reproduce only in living tissue; (2) the poliomyelitis virus has a decided predilection for the tissues of primates; and (3) the monkey is the least expensive and most practical primate to handle on a large scale.

Fig. 55

Business Week, April 16, 1955, p. 38.

forward pass."[45] *Time* hyperbolized that it was the "End of a War," for the vaccine's trial had been covered by the press as massively *as* the end of a major war; and "never before in history had a medical development been big, instantaneous news over a large part of the world."[46] And "big" news it was. Over 1,830,000 children had been studied (440,000 vaccinated; 210,000 dummy; and 1,180,000 controlled) with only one death. Polio was in danger of becoming "as rare as smallpox."[47] (Fig. 54, 55) So impressive was the news in fact, that NBC presented a summary of the test report one hour earlier than mutually agreed upon and was subsequently accused of unfair broadcasting by its competitors.[48]

The exuberance caused by the press, however, was soon tempered by more somber reports: to say that every child was safe was declared patently untrue;[49] only 80-90 percent effective was suggested by Dr. Francis at the University of Michigan's media meeting;[50] and, "restrained applause" greeted the announcement by Dr. Francis.[51] Reports circulated that neither Salk nor O'Connor (NFIP head) had read the report,[52] and Salk stressed that "the most important thing to me is to get back to the lab."[53] In Washington, D.C. Oveta Culp Hobby—Secretary of Health, Education and Welfare—found she had to cancel a press meeting announcing the effectiveness of the vaccine because the Public Health Service had not released its final judgement. Had it been possible Washington would have had a "full dress ceremony."[54]

More confusion was caused by a copyrighted story in the New York *World-Telegram and Sun* which claimed a child who had received the vaccine had indeed contracted polio. And the frustrations resurfaced when "word came from the states of Louisiana, Oklahoma, Virginia and Indiana that the vaccine was not '100 percent safe'."[55]

As more polio cases mounted, defendants of the vaccine were quick to point out that the incidents of polio did "not indicate faulty vaccine" and that the "children may have been well on their way to polio."[56] But few believed such glib reports even when made by the Surgeon General. In fact, Salk himself called for a full investigation amid reports that Albany, New York had banned the vaccine made by Cutter, as had Los Angeles, California. The Detroit *News* attempted to provide guidance for a confused readership:

Q. I am upset about these developments yesterday. Shall I let my child take part in the mass program, or shall I wait?

A. That is an individual decision that must be made by each family, acting on the advice of its family physician.[57]

Hardly an endorsement to take the vaccine, but who could provide a better answer? Curiously the above question appeared next to a jewelry advertisement which read "Don't Take Risks When You Buy Her Diamonds" and before an editorial which suggested the vaccine should be made available to all.[58] There seemed no immediate answer.

Perhaps lyric poet Stephen Spender had less difficulty in answering a question during an interview following an appearance at Wayne State University on the same day that the news broke about new problems with the vaccine. When asked how poetry can aid in solving world problems, Spender suggested that poetry may civilize and humanize societies, but he also conceded that:

Since it takes from 10 to 20 years for a poem to take effect, I can't see how poetry could directly prevent nuclear war.[59]

While waiting for an answer to the polio dilemma, one could find many diversions. The Ten Top Tunes on April 27, 1955 provided adequate escape: "Unchained Melody" (Al Hibbler), "The Breeze and I" (Catherina Velenti), "Don't Be Angry" (Crewcuts), "Most of All" (Don Cornell), "Dance With Me Henry" (Georgia Gibbs), "Opus One" (Mills Brothers), "Dixie Danny" (Laurie Stephens), "Do-Do-Do" (Tony Martin), "My Babe" (Gaylords) and "Heart" (Four Aces). The movies offered even more. One could see *20,000 Leagues Under the Sea* (starring James Mason), *Country Girl* (with Grace Kelly) or *The Barefoot Contessa* (with both Humphrey Bogart and Ava Gardner). Broadway had reached one of its fantastical high points with such productions as *Bus Stop, Can Can, Fanny, Inherit the Wind, Kismet, Tea and Sympathy, Pajama Game* and *Teahouse of the August Moon*. If one were interested in an award winner, *Cat on a Hot Tin Roof*, voted the best play of the 1954-55 season by the New York Drama Critics Circle, was playing to full houses.

However, very little could cover one frightening fact: innoculated children had contracted polio. On April 30, 1955 the count had increased to twenty-six of which California claimed fourteen, Idaho seven, Washington two, and one each from Illinois, Colorado and Georgia. The news became more confusing and fear mounted when papers reported that 300,000 children had already received the Cutter vaccine. No solution was yet found: "Representatives from health agencies emphasized that nothing

The Detroit News

HOME EDITION

THE HOME NEWSPAPER—ESTABLISHED 1873

SEVEN CENTS

Questions, Answers on Vaccine Safety

120 KILLED IN INDO CIVIL WAR

Vaccine Safe, U.S. Authorities Assure Nation

President Blasted by Knowland

GOP Senate Head Might Quit; Hits Change on China

SENATOR KNOWLAND

Boy Admits Bombing at Skillman's

Rebels Battling to Oust Premier

Hundreds of Refugees Caught in Fierce Saigon Cross-fire

Blazes Kill 8 in Chicago; Hunt Firebug

DICK DOES A 'HARRY' AT FUNFEST

Rail Walkout Cripples

Fig. 56

detrimental to health had been found so far in the Cutter vaccine."[60] The Public Health Service indicated that it still supported the innoculation program and "stoutly maintained that its faith in the vaccine had not been shaken"[61] (fig. 56).

By May 1, 1955, the number increased to twenty-nine stricken and California halted all innoculations regardless of the vaccine type. Physicians in New York City were forced to admit they had innoculated at least seventeen adults in violation of the medical profession's own proclaimed self-disciplinary statement that only children would be innoculated. Vaccine used in Chicago was declared safe, but a delay was imposed to permit "more parent observation" since the number of cases was not statistically significant.[62] Photographs of children receiving shots still appeared in newspapers. *Newsweek* summarized the national mood best:

> In a little more than 24 hours, the nation's mood had changed from one of deep relief to one of mass exasperation and frustration stung by fear.[63]
>
> There were reports of "grey markets" developing, and incidents of vaccine theft reported: "A New Orleans mother when told there was no vaccine, pulled two *cc* of it from her pocket: 'Never mind where I got 'em. Just use them for the kids'."[64] Mass hysteria resulted if one case developed in a neighborhood. There were phone calls to governors and congressmen introduced bills to protect their constituents' children. The AFL-CIO called for free public shots in place of commercial channels or private markets. The cost in private hands multiplied thusly: Manufacturers' price $2.70 to $3.12 for 3 *ccs;* drugstores increased it to $3.60, doctors pay $4.20 and the patients paid $6.00.[65]

While President Eisenhower requested that the State Department determine how the anti-polio vaccine production information could be transmitted to other nations,[66] Representative Emanuel Cellar (D—New York) called the voluntary control system being used for the vaccine a "dismal failure" and said the Cutter ban "illustrated in dread fashion the chaos that results from a lack of effective controls."[67] Patrons in movie houses watching *Black Board Jungle*, starring Glenn Ford ("Sway to the Rock and Roll Rhythm of Bill Haley and his Comets as they play Rock Around the Clock"), also learned that a *second* child had died after receiving an anti-polio innoculation. *Time* took an optimistic stance by suggesting that a developing shortage of the vaccine could actually be an advantage in that it would allow for tighter controls against a black market: however, it also reported that the amount of vaccine a doctor received depended, among other variables on how well the salesman liked individual doctors.[64]

By May 3, the emotional responses were dissipating and more controlled reactions were manifest: An editor called for calm by stating "it is not a time for hysteria but a time to trust public health authorities"; however, one citizen expressed his bewilderment in this unique statement:

> If we are so ill-prepared as a nation to deal with the arrival of a benison like the Salk vaccine, heaven help us if we should have to deal with a national catastrophe.[69]

Parents' willingness to cast "their fortunes with the findings of science" forecasts a "speedier success against polio," wrote one editor.[70] However, such an overwhelming number of news stories which cast doubt on the vaccine's safety buried such editorialized optimism in a mire of reports which told how "officials were confused," how "confusion gripped the nation," how "the government had suddenly reversed itself"—and in ominous photographs which showed stricken children lying in hospital beds.[71] Holland decided to delay vaccine use, Sweden called off its vaccine program, and Austria decided on a German vaccine. The count had risen to fifty-two victims.

Questioning of the vaccine's safety became daily fare for news readers:

> If vaccine caused any of the 52 polio attacks among those vaccinated how could it happen? A number of explanations have been offered. Maybe none of them is right.[72]

"Why polio developed in some children from the same bottle" puzzled researchers, doctors, health officials and vaccine manufacturers. *Life* presented a photo of a stricken child from Idaho and of a thirteen-month-old child in Colorado.[73] Calls such as "no time for panic" seemed hollow indeed especially when juxtaposed to a story of the six-year-old San Diego boy whose case gave rise to the question as to whether he "caught the disease from defective vaccine" or whether he had the disease *before* he was innoculated.[74] "The vaccine itself may involve some risk," claimed *U.S. News and World Report,* and "Salk shots, contrary to what many thought, promise no sudden, sure end to polio."[75] No one seemed certain on the safety question, no one seemed even to be able to provide a confident course of action.

Certainly the Office of the Surgeon General helped little with its contradictory actions. On May 16, 1955, *Newsweek* reported that Dr. Leonard A. Scheele, Surgeon General, recommended postponing the vaccination program; yet at the same time, his office claimed "We have every faith that this vaccine ... is both safe and effective."[76] Amid inferential criticisms of both Jonas Salk and the NFIP, the contradictions continued from that office: only one in 68,000 victims died from polio *yet* polio was declared "a national problem"; there were flaws in the virus death rate, yet researchers' knowledge and intuition were praised; and, manufacturers had not given the government full information on production problems, *yet* the manufacturers "had come through with flying colors."[77] It seemed the government was attempting to place some responsibility for the problem on just about any organization which had anything at all to do with the vaccine. Producers were accused of not fitting their production to Salk's original concept of the vaccine and the NFIP was called down for not providing enough statistical data to establish adequate manufacturing procedures.

Perhaps the national anger was best crystallized in the following, amid "bitter disappointment" and "utter chaos":

> The nationwide program of vaccination against polio, so eagerly awaited for so many years and greeted with clarion calls of hope, ground this week to a screeching halt.[78]

President Eisenhower attempted to bolster public confidence in the vaccine with a news conference that the Washington *Post* called "a miracle of confusion." "President Set to Take Over Salk Vaccine, if Necessary" and "Ike Promises Free Shots for All 5-9" were typical contradictory headlines in May 1955. On May 4, Eisenhower stated the national program would continue, but by May 14, it had been stopped, albeit temporarily. Cities began asking for more details on the vaccine's safety, but no adequate response was forthcoming. Expressions of confidence were challenged by charges of doubt, especially since the administration was still under attack from those who believed no adequate program had been implemented for equitable and effective distribution of the vaccine. The best the public could be given was opaque confidences, such as the following weak reassurance:

> President Eisenhower is justified in his confidence that here too [Detroit] the program would be adhered to. Public opinion, he said, will see to that.[79]

But public opinion had been shaken by the mounting confusion, and by other puzzling qualifications: "You don't have any choice. It's got to be right," and "as long as the manufacturing rules are scrupulously obeyed by scrupulous men, then the vaccine must be safe."[80] The fairness of the program had already been questioned in terms of equitable distribution, now the manufacturers' "scruples" were being highlighted. In addition, the "vaccine grey market" was again being discussed in the media.

According to one *Time* report, eighty packages of the vaccine had been distributed "as largesse to employees of the drug houses and their friends or relatives."[81] In addition, there were charges that private doctors had distributed it "as they pleased with no regard to priorities."[82] *The New Republic* claimed that over fifty national welfare groups were objecting to the voluntary distribution system since vaccine was not getting to the 5—9 age children as Eisenhower had promised. In fact, the editors called for a federal commission for allocation since the vaccine was "going to friends of the drug manufacturers, doctors' children, one-year-olds and adults."[83] *Business Week*, more conservative and more restrained, also demonstrated concern over the possibility of the "grey market" becoming a "black market":

> In order to make sure that no vaccine is being diverted into black market channels, the Drug and Food Administration plans to hire pharmacy school students to help FDA inspectors check wholesale and retail drug stores.[84]

As if the Salk vaccine were not sufficiently embroiled in enough continuing controversy, yet another unresolved question resurfaced: Was the Salk vaccine indeed the best vaccine? Salk supporters agreed that the "killed virus was better than live virus" and that there was "less risk with dead virus."[85] Proponents of the live vaccine claimed the following: "A vaccine made from living polio virus, as distinguished from the 'killed' virus of the Salk vaccine, is the eventual answer to the paralysis problem."[86] They claimed the 'killed' vaccine was not "a panacea from a magician's hat and that the national program was an "opportunity for some bureaucrats [sic] to demonstrate the glorious material instincts of 'big brother'."[87] More confusion, more doubt, more concern and still no solution, still no guaranteed safe vaccine, still no adequately equitable distribution, and still fifty-two victims of polio after innoculation:

It is the heaviest investment ever made, in money or faith, in the belief that youth will sustain such a gigantic enterprise.[88]

This quote also faced the reader early in May 1955: however, it was not related to the vaccine controversy—it had to do with the planned opening of Disneyland, an announced opening which accompanied announced closings of polio clinics nationwide. Jonas Salk was honored by Roosevelt University, however the following has to do with yet a different award:

We see no reason why the public should strain its powers of aesthetic appreciation or its sense of moral values to understand this particular award.[89]

The rebuke is directed at the Pulitzer Prize being awarded to Tennessee Williams' *Cat on a Hot Tin Roof*. Press photographs of family fun at Sun Valley, by the way of the Union Pacific and stories of a 100 percent Industrial Tax exemption by Puerto Rico for new industries found their way between the vaccine stories. One of *McCall's* featured articles asked the burning question: "Are Diet Quacks Ruining America's Health?" Another problem was solved by the following medical definition of a transvestite: "One who had a morbid desire to dress in the clothing of the opposite sex."[90] Such individuals were classified as "sex deviates"; however, one such was declared "cured" when he "got a crew cut," "took a girl friend home to mama" and, in addition, "applied for a branch of the armed services."[91]

One of Sid Caesar's parodies struck two anathemas at the same time: hair styles and rock and roll music. In a mimic of the Crew cuts, Caesar satirized rock 'n' roll with a creation called "Three Haircuts," and played music with a screaming and honking background. And the assault on rock music did not stop with Caesar's television skits for a new record—a proclaimed best seller—had arrived, "The House of Blue Lights" (Mercury) with Chuck Miller: "A boogie-woogie in uptempo, with some nonsense words about boogie-woogie. Does it herald the decline of rock 'n' roll?"[93]

A far more important question, however, was about to be answered for "a definite relationship between innoculation with Cutter Laboratory vaccine and polio had been discovered."[94] As early as May 9, 1955, questions were being asked if "a laboratory slip up in the job of killing the virus" could be accountable for the

MEDICINE

Halt!

The nationwide program of vaccination against polio, so eagerly awaited for so many years, so recently greeted with clarion calls of hope, ground this week to a sickening halt. The U.S. Public Health Service recommended (and all states were virtually certain to comply) that use of the Salk vaccine be postponed until it can "reappraise" the vaccine now on hand. This includes 1) vaccine shipped to public authorities and now in their refrigerators (enough for 4,000,000 or more shots), plus 2) a similar amount still in the manufacturers' vats.

Fig. 57

Time, May 16, 1955, p. 57.

THE POLIO STORY:

Whose Fault?

The tragic near-breakdown of the anti-polio inoculation program erupted last week into sharp criticisms and obvious efforts to shift the blame.

Surgeon General Leonard A. Scheele of the U.S. Public Health Service gave his side of the story in a lengthy report to Welfare Secretary Oveta Culp Hobby. In it, Scheele inferentially criticized both Dr. Jonas E. Salk, developer of the vaccine, and the National Foundation for Infantile Paralysis. Scheele said manufacturers had found that the process of rendering harmless the live virus in the vaccine "does not always follow the original concept" of Salk. And he implied that the National Foundation had failed to give the Federal government sufficient statistical data to set up "adequate" procedures to determine manufacturers' level of performance.

Salk was silent, but Basil O'Connor, president of the foundation, fired off an indignant statement. He said: "The report obviously gives only the Public Health Service version of the vaccine situation to date. When the public has heard Dr. Jonas E. Salk's version, it will be able to form an intelligent conclusion as to what actually has been transpiring."

Fig. 58

Newsweek, May 16, 1955, p. 49.

MEDICINE

Premature & Crippled

The nation's top polio experts and health authorities got around last week to telling the public what had gone wrong, and how and why, with the grandiose plan to inoculate tens of millions of children with Salk polio vaccine in 1955. For the most part, they confirmed what critics have suspected for a long time: it was a mistake to try the whizbang jump in little more than a year from laboratory production of the vaccine to manufacturing in tank-car volume. Many vital facts simply were not known when the leap to factory-scale production was made. In the light of this new knowledge, they implied, not only the methods of making and testing the vaccine but also the vaccine itself will be drastically altered as soon as possible.

Surgeon General Leonard A. Scheele of the U.S. Public Health Service and his deputy, James Shannon, picked a polio symposium at the American Medical Association's convention in Atlantic City as the forum for their first report. Then Dr. Scheele released a 163-page "white paper" for President Eisenhower.

Fig. 59

Time, June 20, 1955, p. 45.

"rash of polio and dead children."[95] Why had some doctors in California decided they "wouldn't give it to their own kids?"[96] Finally, distribution of the Cutter vaccine was stopped for indeed live virus had contaminated the vaccine (fig. 57, 58, 59).

But who or what was responsible for the "Vaccine Snafu":

> In retrospect, a good deal of the blame of the vaccine snafu also went to the National Foundation, which, with years of publicity, had built up the danger of polio out of proportion to its actual incident, and had rushed into vaccination this year with patently insufficient preparation.[97]

As the cases mounted to eighty-nine reported victims, more agreement was reached with "last years' skeptics that it would have been better to give over 1955 to further tests."[98] *The New Republic* criticized the manufacturing process by indicating that "production had been faster than testing" and that this was able to happen because the "public was aroused by high pressure publicity and ceremony"; going further, the editors claimed this was an expected result from "over-dramatizing every new scientific achievement."[99] The government was forced to concur with this appraisal, for the Surgeon General's office agreed that "spot checks" were done based on prior approval of commercial laboratory techniques—this was explained to the public as being done on the basis of approval protocol.[100] "It was a mistake to try the whiz-bang jump in little more than a year from laboratory production of the vaccine to manufacturing in tank car volumes," *Time* readers were informed "you cannot make viruses meet deadlines."[101]

Once the problem was identified, the press reflected a call for new procedures and testing: *Business Week* called for raised standards "to insure greater safety." Laboratory standards in universities were appropriate for research but mass produced procedures were needed if the Salk vaccine program were to continue.[102] Others viewed the vaccine's problem in a more optimistic light. Production delays were termed beneficial because "from the delay science has gained new knowledge, new safeguards.'[103] In fact, Under Secretary of State Nelson Rockefeller saw an opportunity "to sweep up some of the dust caused by the polio row."[104] Some claimed the contaminated vaccine proved that pure vaccine was "safe, generally"; and backed their opinion with statistics showing that of 6,000 innoculated only 94 were stricken and only 73 of these were paralytic. Such figures compared favorably to previous polio incidents and paralysis ratios.[105] Only four percent had minor reactions; and accidents which occurred from the vaccine "were the exception and not the rule."[106]

Finally, the report on the Cutter vaccine that most were waiting for was released by the Public Health Service. Six lots of vaccine produced 103 cases of polio due to ineffective live virus and the PHS claimed the following: "The exact reasons for the presence of...virus...could not be found" and there was "no evidence that anything was done by Cutter to short cut or be sloppy deliberately."[107] Cutter Laboratories, which was said to have claimed in a report to stockholders "to be up to our ears in poliomyelitis vaccine production,"[108] was understandably relieved. In fact the degree of relief can probably be seen by the reported reaction of Dr. Cutter: "A weight off my shoulders. I'm going home to work on my camellias."[109]

Autumn 1955: Anticlimax and Business as Usual

When viewed relative to the first half of 1955, one finds very little overt excitement or controversy concerning the Salk vaccine. Of course, there were still significant claims, counterclaims and new discoveries, but generally speaking the vaccine was in use, it was considered effective and safe and most of the turmoil had vaporized. (Figures 60, 61, 62, 63) The following perhaps provides a representative sample of the kinds of stories people were reading.

Editors who were politically oriented generally wrote in one of two directions: more government control of the innoculation program *or* less government interference. *The New Republic* argued that in principle "some Americans seem to regard socialized medicine as a worse calamity than the deaths of a few children" and that the Administrative policies represented protectionism for members of the American Medical Association.[110] *Business Week* postulated that the nation has escaped a polio epidemic despite a "government slip" which consisted of the "terrifying fact that the government wasn't sure of its own standards, (as) reported by a USPHS report."[111]

> It appears plain that medical and public health authorities failed to take account of the dangers that always accompany a new, powerful and comparatively untested drug.[112]

Those who believed in the positive advancement of science through technological means envisioned the anti-polio vaccine as the beginning of an elimination to all viruses, with the Salk vaccine being the "first, brilliant triumph in the all-out assault":

Here, for the most part, unnoticed by the public, is a medical revolution of the same order of magnitude as that which followed the discovery of penicillin.[113]

Many reports in the media contended that polio "as a major menace had been licked since the results for the six months preceeding October 1955 were better than hoped for." Further, the question of the vaccine had been settled to the satisfaction of most and despite continual shortages of the vaccine the general contention, in the media at least, was the "Salk vaccine is safe!"[114] Revised testing and production procedures cleared the vaccine from any association with polio, thereby reinforcing the technocrat's contention that the "Science" of viral destruction was indeed intact; the problem, therefore, was not viewed as technical but one of simple production errors.

Despite such apparent scientific success, there were still detractors. The British Health Authorities considered the vaccine too dangerous and called for less virulent strains to be used. Said the Director of Public Health: "I do not see how any vaccine by Salk's method can be guaranteed to be safe."[115] In fact, the British opposition was so strongly supported that a modified Salk vaccine, made from the milder Brunhilde strain, was employed, with the belief that any mistakes were less able to cause paralysis of the serious sort.[116] Unfortunately, the early British innoculation program was even less successful than the American one. In addition, as late as September 1955 doubts continued to linger about the vaccine in the United States. Reports indicated that some laboratories "have yet to get any batches of their vaccine under the higher safety standards that the PHS set in May."[117]

In effect, the Salk vaccine innoculation program had experienced a multitude of related and unrelated bruises. Curiously, the vaccine was proven both safe and unsafe, the ominous warning of Walter Winchell had been verified but at the same time the promise of polio control had been realized. The greatest medical experiment in United States history had been conducted; the results were positive and the future hopeful. Quite naturally, the media reported as much to the American public. However, the anti-polio vaccine had yet another chapter—one which would stretch over at least the next three years.

1956-1958 "Walk With Salk"

There was, in fact, a great fallout which resulted from the

Fig. 60
May 16, 1955, p. 84.

Fig. 61

U.S. News and World Report, April 22, 1955, p. 97.

Fig. 62

Such borrowing of prestige permeated the advertising world, suggesting that relationship between the advances in medical science and one's own business.

Fig. 63
Time, June 29, 1955, p. 38.

1955 mass innoculations; and, generally seven areas monopolized the vaccine 1956-1958 polio news—the publicity question, the British program, general information, continued supply problems, the Chicago epidemic, lawsuits and the "live vs. dead" vaccine argument.

The publicity controversy was still heated. In addition to the 1955 charges of showmanship and "Hollywoodisms," more direct accusations surfaced, the most ironic, seemingly, that the medical profession was attempting to protect its members. One article in the *Saturday Review* addressed itself in no uncertain terms to the publicity charges: People were reading in their newspapers the favorable arguments for the vaccine being published by the NFIP. The unfavorable news was being played down.[118] If true, such accusations pointed to the worst type of public manipulation through censorship; if untrue, a certain percentage of readers would believe the indictment as fact, thereby negatively affecting the vaccine program. The manipulation contention was further reinforced by the *Journal of the American Medical Association* which reported that the *Bulletin of the American Association of Public Health Physicians* claimed "the public is being urged from all directions, except that of the practicing physician, to get their innoculations immediately."[119] Indeed, part of the American Medical Association's concern resulted from the fear of lawsuits which would inevitably result from alleged malpractice. One father in fact did file suit for $300,000 claiming his son had contracted polio from the innoculations.[120] And Cutter Laboratories was faced with thirty-three damage suits for millions of dollars. Melvin Belli, attorney for the plaintiffs, maintained that the contaminated vaccine was a violation of implied warranty, and one jury (eight women, four men) had ruled eleven-to-one to award one family $147,300 for two affected children; however, the company which produced the vaccine was absolved from any recklessness.[121]

News reports of August 1957 contained disturbing news of the highest order: "U.S. vaccine (was) being smuggled and sold on the Black Market" in England because the British program against poliomyelitis had been "a dismal flop." Statistics relayed that the stricken rate was 87 per 55,000, very high indeed. Such a disaster resulted in part from what *Time* called "Pride above Polio."[122] The U.S. rate was 6.3 per 100,000 for innoculated and 29.2 per 100,000 for those not innoculated.

Luckily, the vaccine problems in the United States were that of periodic shortages and of individuals refusing to have innoculations when the vaccine was in fact available. News stories claimed there were seventeen million unused doses on producers' shelves;[123] and although there was enough vaccine for "every man, woman and child," less than fifty percent of those in the most susceptible age group had been innoculated.[124] Fifty five million Americans under forty, the polio susceptible age group, were said not to be immunized and walked "the street unaware or unconcerned that they have no polio protecting shots."[125] Such news in January of 1959 seemed especially ironic when juxtaposed to the 1957 shortages. Reports had indicated that by March 1957, a 26,000,000 backlog had been reduced to "almost nothing," yet still 27,000 doses had to be destroyed because of "public apathy."[126] In April 1957, *Time* informed its readers that there were only 3,200,000 doses for fifty-eight million people.[127] Yet for three years, vaccine proponents continued to maintain that "the long siege is ended" and that the Salk vaccine could indeed "lick polio."[128]

The Chicago polio outbreak earlier in 1956 had paid grim testimony to types of supply and distribution problems which existed. One frightening statistic from this epidemic demonstrated that of five hundred victims only sixty-one had gotten innoculated. The disaster which occurred in the "slums" was explained away by the most glib expediency:

> Authorities noted that the hotbed of the Chicago outbreak has been in a slum area where the inhabitants are either foreign born or of foreign extraction. They apparently have not yet been sufficiently assimilated into the U.S. pattern of living to take advantage of opportunities to get the vaccine even through free clinics.[129]

The impact of such subjective reporting was not lost on doctors in North Carolina, the next to last state to begin innoculations, with Idaho being the last:

> The medical profession in North Carolina hoped to avoid the situation in Chicago, where polio has been attacking small children who did not get vaccinated because of parents' indifference or lack of clinics.[130]

In fact, to avoid "the situation" doctors in Greenboro, North Carolina, apparently seduced teenagers with the following: "Walk with Salk so you can rock and roll."[131]

The Salk vaccine survived the myriad of complications it was forced to confront, yet there was one opponent to which it eventually had to yield: the live vaccine. The argument of live vs. dead vaccine had raged in the press and in professional journals for many years, but it seems the confidence expressed in news reports from about March of 1956 heralded a new approach to controlling viruses:

> The day is coming when children will drink milk shakes or swallow capsules of live polio virus safely and get protection against the disease.[132]

Even the more restrained reports illustrated the inevitability of technical change inspired by increased scientific knowledge: Researchers believed "the future of live virus vaccinations is 'bright from the point of view of scientific developments'."[133] In fact, the oral vaccine had been tested among one hundred and thirty prisoner volunteers who received a single oral dose of the Sabin vaccine; however, more data was needed.[134] Tests were conducted on monkeys and chimpanzees and the live vaccine was being touted as an answer in a "pill containing something called 'Sabin vaccine'. "[13] Since 1940, the oral vaccine experiments had used only about a million dollars and was recognized as a good investment by many. In fact, Sabin himself argued that "the only way to wipe out polio is to make the human intestinal tract immune to reinfection."[136] A more objective analysis appeared in *Scientific American* and reached the same conclusion:

> The immunity conferred by dead virus does not immunize the intestinal tract. Live virus does, therefore polio cannot be passed from person to person in a disease free population as it can with a killed virus.[137]

There was a massive innoculation program using live virus carried on in Africa in an area designated as the "Congo Epidemic Area" where 244,596 individuals were innoculated with "no sickness reported."[138] These results were said to be identical with six hundred who were innoculated in the United States. *Time* reported that 215,504 men, women, and children were innoculated in the Ruzizi River area between Ruanda and the Congo with only two negative reactions from children who failed to create adequate antibodies.[139] However, the live vaccine was still judged unsafe for general use in the United States.

It was usually acknowledged by the press that the Sabin vaccine was a logical follow-up to the Salk vaccine; whether, in fact, this is indeed so is debatable for the vaccines may be so conceptually different as to defy any cause-effect paradigm. Some were quick to concede outright that based on tests to date (December 1958) "the live polio vaccine is a better overall immunizer than the Salk vaccine."[140] The more popular view seemed that the attenuated live vaccine "could deliver the knockout blow to the disease, sent reeling by Jonas Salk's 'dead' vaccine."[141]

Medical progress in the name of science is no disgrace, indeed it is desirable and inevitable for mutability is the key to development in technological societies. Even insignificant creations take on unwarranted proportions under the guise of achievement—a 1956 homemaker who switches on an automatic laundry dryer in perfect safety (supposedly happening every fifteen seconds somewhere in the United States) or a new plastic golf club whose precision never varies thereby never causing a hook or a slice on the golf course.

There are, however, some nagging questions about the Salk vaccine which the press highlighted without fully answering. Was the dead vaccine the best vaccine? Was it fully tested? Indeed, did the rush to produce co-opt the need to test? Was the dead vaccine necessary at all, or should the live vaccine have been perfected? Was the Administration equitable and efficient? Were the production safeguards adequate? Did the polio publicity create an hysteria in which logical judgments could not be made? Was there a "grey market" or a "black market"? By 1959 the controversies had not yet been adequately settled, at least not in or by the press.

Notes

[1]"Polio Vaccine Effective," *Science News Letter*, April 11, 1953, 63:211.

[2]*Science News Letter,* 24 October 1953, 64:259.

[3]"Mass Polio Test," *Newsweek*, 23 November 1953, 42:62.

[4]"Polio Vaccine Progress" in "Science and the Citizen," *Scientific American,* December 1953, 189:52-53.

[5]"Question Mark Vaccine," *Detroit Free Press*, 27 April 1954, 32: back page.

[6]"Polio Vaccine Trials Evaluation Planned," *Science News Letter,* 20 February 1954, 65:12.

[7]*Op. cit.*

[8]"Polio Vaccine Trial," *Science News Letter,* 13 February 1954, 65:100.

[9]"Polio: At Last, The End of The Crippler," *Newsweek,* 22 March 1954, 43:94.

[10]"Is Polio Vaccine Safe?" *U.S. News and World Report,* 26 March 1954, 36:35-36.

[11]"Live Virus Found, Says Commentator," *Ann Arbor News,* 5 April 1954, 2:5.

[12]"Health Units Call Polio Vaccine Safe," *New York Times,* 5 April 1954, 27:7.

[13]"Vaccine for Polio Affirmed as Safe," *New York Times,* 6 April 1954, 33:1.

[14]"Vaccine Safety," *Time,* 12 April 1954, 63:15.

[15]"Polio Vaccine Safe," *Science News Letter,* 24 April 1954, 65:264.

[16]"Five Take Fake Test for Polio Serum," *New York Times,* 8 April 1954, 30:4.

[17]"County in Michigan Cancels Polio Tests," *New York Times,* 10 April 1954, 13:1.

QPW"Medical Society Delays Vaccine OK," *Ann Arbor News,* 5 April 1954, 1:8.

[19]"The People Are Not Too Dumb To Know," *Detroit Free Press,* 10 April 1954, 8: editorial.

[20]"269,000 Needles," *Time,* 3 May 1954, 63:68.
[21]"Polio Pioneers," *Time,* 26 April 1954, 63:53.
[22]"Vaccination Day," *Newsweek,* 3 May 1954, 43:84.
[23]*New York Times,* 9 April 1954, 21:6, book advertisement.
[24]"52 Tons...," *New York Times,* 8 April 1954.
[25]Taubman, Howard, "Issac Stern Beethoven Concert," *New York Times,* 8 January 1954.
[26]*New York Times,* 7 April 1954, 22:3.
[27]*Chicago Daily Tribune,* 7 April 1954, 1:4.
[28]Couglan, Robert, "Tracking the Killer," *Life,* 13 August 1954, 120-121.
[29]"Polio Gamble," *Time,* 1 November, 1954, 64:77.
[30]"Polio: Free Shots for 9 Million," *Newsweek,* 1 November 1954, 44:62.
[31]"News from Salk," *Time,* 20 September 1954, 64:88.
[32]"Ballet Theatre Opens at Metropolitan," *New York Times,* 13 April 1955, 34:1.
[33]"Movie Outlook Bright," *New York Times,* 14 April 1955, 34:1.
[34]"Citation for Dr. Salk," *New York Times,* 24 April 1955, 10E:3.
[35]This attitude was generally expressed in many sources, in part as a reaffirmation of the capitalistic ethic.
[36]"U.S. Gets Set for Polio Vaccine," *Life,* 11 April 1955, 38:35-9.
[37]*Ibid.,* p. 38.
[38]*Ibid.*
[39]"Dawn of a New Medical Day," *New York Times,* 13 April 1955, 28:1, editorial.
[40]Lawrence, David, "Man's Miracle?" *U.S. News and World Report,* 22 April 1955, 38:156.
[41]"Mrs. Roosevelt Pleased by Salk Test Outcome," *New York Times,* 12 April 1955.
[42]"New Polio Vaccine," *U.S. News and World Report,* 22 April 1955, 38:44.
[43]"Fanfare Ushers Verdict on Tests," *New York Times,* 13 April 1955, 1:8.
[44]"Vaccine for Polio: The Problem Now is Production," *Business Week,* 16 April 1955, 136-140.
[45]*Ibid.,* p. 140.
[46]"What Is It?" *Time,* 25 April 1955, 65:51-52.
[47]"It Works," *Time,* 25 April 1955, 65:50.
[48]"Release Broken by NBC on Polio," *New York Times,* 13 April 1955, 40:4.
[49]"Vaccine Verdict," *Time,* 11 April 1955, 65:74.
[50]"Polio Protection for Life," *Science News Letter,* 23 April 1955, 67:259.
[51]"Fanfare... *Time.*
[52]In fairness to Dr. Jonas Salk, it need be stressed that throughout the media ordeal, he maintained the highest standards of ethics relative to experimentation and research. In fact, he constantly and consistently is reported to have cautioned against reaching conclusions too rapidly.
[53]"What Is It... *Time.*
[54]"6 Vaccine Makers Get U.S. Licences," *New York Times,* 13 April 1955, 1:7.
[55]"That Polio Vaccine," *Newsweek,* 11 April 1955, 45:74.
[56]Furman, Bess, "One Firms Vaccine Barred," *New York Times,* 23 April 1955. 1:2.
[57]"Questions, Answers on Vaccine Safety," *Detroit News,* 28 April 1955, 1:198.
[58]*Ibid.,* p. 42:1.
[59]Rose, Judith, "Drugged English Poet Refutes Stereotype of Bookish Bard," *Detroit News,* 20 April 1955, 37:1-3.
[60]Kaplan, Morris, "300,000 Got Cutter Shots," *New York Times,* 30 April 1955, 1:2.
[61]"Trouble for the Salk Vaccine," *Business Week,* 30 April 1955, p. 26.
[62]"Delay Oak Park Polio Shots One Week," *Chicago Sunday Tribune,* 1 May 1955, p. 1:5-6.
[63]"The Polio Scramble," *Newsweek,* 2 May 1955, 47:79-81.
[64]*Ibid.,* p. 79.
[65]*Ibid.*
[66]"The Salk Polio Findings," *Commonweal,* 29 April 1955, 62:94.
[67]"Hits Hobby's Plans for Salk Vaccine Sales," *Chicago Tribune,* 2 May 1955, 30:2, editorial page.
[68]"Where is the Vaccine," *Time,* 2 May 1955, 65:92.
[69]"The Salk Vaccine," and "Letter to the Editor," *New York Times, e May 1955, 30:2, editorial page.*
[70]"Vaccines and Faith," *Detroit News,* 3 May 1955, 24:1, editorial page.
[71]"Vaccine Curb Stirs Nation," *Chicago Tribune,* 8 May 1955, 1:8.
[72]Marlow, James, "Vaccine Not Yet Definitely to Blame for Polio Attacks," *Ann Arbor News,* 9 May 1955, 4:4-5.

[73]"Sudden Hitch in a Program," *Life,* 9 May 1955, 38:38-9.
[74]"Vaccine Crisis," *Time,* 9 May 1955, 65:49.
[75]"Truth About the Polio Scare," *U.S. News and World Report,* 13 May 1955, 38:25.
[76]"The Polio Story: Million—and 52," *Newsweek,* 16 May 1955, p. 49.
[77]"The Polio Story: Whose Fault?" *Newsweek,* 20 June 1955, 45:22.
[78]"Halt?", *Time,* 16 May 1955, pp. 57-8.

[79]The Promises Free Shots for All 5-9," *Detroit News,* 5 May 1955, 38:2.
[80]"Safeguards for the Polio Shots," *Life,* 16 May 1955, 38:139-140.
[81]"Vaccine Grey Market," *Time,* 9 May 1955, 65:50.
[82]*Ibid.*
[83]"The Vaccine—Whose Health and Whose Welfare," *The New Republic,* 9 May 1955, 132-3.
[84]"Now that concern over the safety of the vaccine has subsided..., *Business Week,* 4 June 1955, p. 72.
[85]Salt, Jonas, "Vaccine for Poliomyelitis," *Scientific American,* April 1955, 192:42-44.
[86]"Living Virus Vaccine Held Key to Polio," *Chicago Tribune,* 3 May 1955, editorial page.
[87]*Ibid.,* "Voice of the People."
[88]"Disneyland Surprises to be Unveiled in July," *Detroit News,* 8 May 1955, p. 168.
[89]"A Strange Award," *Christian Science Monitor,* 14 May 1955, editorial, 1-2.
[90]"Medical Definition of Transvestite," *New YOrk Times,* 27 June 1955, p. 43.
[91]*Ibid.,* p. 47.
[92]*Time,* 21 June 1955, p. 41.
[93]*Ibid.*
[94]*Cf.* "Vaccine Crisis" *Time,* p. 49.
[96]*Ibid.,* p. 50.
[97]"Vaccine Snafu," *Time,* 30 May 1955, p. 40.
[98]"Vaccine Evidence," *Time,* 23 May 1955, p. 52.
[99]"Salk Shots: Government Reluctancy Acts, *The New Republic,* 16 May 1955, 132:3.
[100]*Cf.* "Vaccine Not Yet...", *Ann Arbor News.*
[101]"Premature and Crippled," *Time,* 20 June 1955, pp. 45-46.
[102]"Tests Mean New Salk Delays," *Business Week,* 25 May 1955, p. 32; "Vaccine Safety," *Time,* 4 July 1955, 66, p. 60.
[103]"Questions Without Answers," *Business Week,* 25 May 1955, pp. 45-46.
[104]*Ibid.*
[105]"Clearing Up The Polio Scare," *U.S. News,* 3 June 1955, 38:30-32.
[106]"Effective Safe," from "Science and the Citizen," *Scientific American,* JUne 1955, 192-46; and "Nearly Eight Million Get Polio Vaccine Safely," *Science News Letter,* 13 August 1955, 58-104.
[107]"Finally, the Report," *Newsweek,* Sept. 1955, 46:65.
[108]"Trouble at the Plant," *Time,* 20 June 1955, p. 84.
[109]"PHS Report of the Cutter Vaccine," *Time,* 5 Sept. 1955, 66:36.
[111]"Salk Vaccine," *Business Week,* 3 Sept. 1955, p. 32.
[112]"Victory Over Polio" from "The Trend," *Business Week,* 26 Nov. 1955, p. 196.
[113]"Science Moves in on Virus," *Life,* 20 June 1955, 38:122-4.
[114]"Plenty of Vaccine—When," *U.S. News,* 16 Sept. 1955, 39:24-5.
[115]"Vaccine Dissent," *Time,* 25 July 1955, p. 35.
[116]"British Salk," *Time,* 30 Jan. 1956, pp. 67-8.
[117]"Doubt Salk Vaccine Program Can Meet Christmas Target," *Business Week,* 24 Sept. 1955, p. 128.
[118]"A Report of the Salk Vaccine," *Saturday Review,* 24 March 1956, 39:64-5.
[119]*Ibid.*
[120]"Vaccine and the Law," *Time,* 27 Jan. 1957, 69:83.
[121]"Verdict on Vaccine," *Newsweek,* 27 Jan. 1958, 51:54.
[122]"Pride Above Polio," *Time,* 26 August 1957, 70:77-78.
[123]"If You're Under 35," *Newsweek,* 10 Dec. 1956, 48:70.
[124]"Bog Down in Polio," *Newsweek,* 20 August 1956, 48:54.
[125]"Get that Fourth Shot," *Newsweek,* Jan. 1959, p. 80.
[126]"What Happened to All the Polio Vaccine?" *Science News Letter,* 30 March 1957, pp. 71-80.
[127]"Sorry, No Vaccine," *Time,* 8 April 1957, 69:42.
[128]"The Long Siege is Ended," *U.S. News,* 26 July 1957, 43:72-3.
[129]"All Ages Now Can Get Polio Vaccine," *Science News Letter,* 4 August 1956, 70:72.

[130]"North Carolina Doctors Lead in Polio Fight," *Science News Letter*, 31 March 1956, 69:201.

[131]"Walk With Salk," *Time*, 6 August 1956, 67:37.

[132]"See Bright Fugure for Live Polio Virus Vaccine," *Science News Letter*, 31 March 1956, 69:201.

[133]"Live Polio Vaccine" from "Science for the Citizen," *Scientific American*, May 1956, 194:60.

[134]"Vaccination by Mouth, *Time*, 15 Oct. 1956, 68:64.

[135]"New Polio Serum Answers for Parents," *U.S. News*, 19 Oct. 1956, 41:64-6.

[136]"Need Live Virus to Stop Spread of Polio," *Science News Letter*, 20 July 1957, 72:40.

[137]"Live-Virus Polio Vaccine," from "Science for the Citizen," *Scientific American*, Sept. 1957, 192:112.

[138]"Live Polio Vaccine Aids in Congo Epidemic Area " *Science News Letter*, 9 August 1958, 74:94.

[139]"Live Virus in the Jungle," *Time*, 11 August 1958, 72:32.

[140]"Virus Vaccine Tested," *Science News Letter*, 6 Dec. 1958, 74:355.

[141]"Researchers Ready to Start Tests on Live Polio Vaccines," *Business Week*, 9 Nov. 1957, p. 187.

Afterword
Suggestions for Further Study

In retrospect the press in part reflects the cultural tenor connected with each technological event discussed in this book. How effective the press was in influencing decisions toward one course or another on each issue is debatable; and, in essence, the influence of the press may never be fully measured. Yet the question of the relation of the press to technological change is still viable; and how the press attempts to influence, support or negate technological change is a valid question in a free society.

Perhaps the following questions provide suggestions for continuing research.

First, does the press have the right to influence its readers on technological issues? Opinions vary on this issue. In contemporary America, one must consider the fact that today's newspapers number far fewer than the number that existed in previous years. Somehow advocacy reporting seems somewhat more tolerable in a city with a dozen dailies and numerous weeklies than the same advocacy in a city with a few papers and limited numbers of national weeklies.

Two, does the press, except in editorial situations clearly labeled as such, report objectively on technological events? This is particularly important when writing on technological issues since the issue itself appears objective. In other words, the technological subject is surrounded with a cultural aura of objectivity which can, in some cases, conceal subjective leanings. One expects subjectivity on the sport pages and the editorial pages of newspapers—yet one expects objectivity in technological and scientific news. Does subjectivity cloaked in an objective aura mislead the reader?

A third question arises relative to the reporting procedure itself. Do news reporters understand the technological issue clearly enough to report comprehensively? The examples cited in this book illustrate that more is involved in technological change than changing technologies. This question can and need be extended to editors as well.

Fourth, how closely should the press work with those calling for technological change? It seems, from the examples in this book, and from others, that positive technological change demands immediate space while negative news of technology is most often ignored or buried on the back pages. The exceptions to this general rule are sensational stories or coverage of highly controversial technologies.

A fifth question needs to be addressed to the rapid communication of advanced technological cultures. Does the rapidity of television and radio somehow give instant recognition and instant sanction to technological change? Of course little can be done to alter the technology reported upon; however, and very importantly, the accelerated pace of technological reporting in itself accelerates change by preparing the view for the *next* change, the one after that, and so forth.

Six, how much influence does the press indeed exert? Except for those in Washington and New York City, the Panama Canal issue at the turn of the century was somewhat removed and vague. The emotional tenor of some reporting probably did not greatly affect supporters and detractors of one position or the other. Yet the effect of the press on the populous must be judged somewhat ineffectual. For most Americans at the turn of the century Panama was too far; besides, there were immediate local problems commanding more attention. The reporting on the V-weapons and the atomic bomb seemed more immediate although great geographic distances were involved. On August 7, 1945 Hiroshima was as near as any newstand. And, as mentioned in the previous textual material, citizens on the East Coast of the United States were haunted by the V-weapons in a real, yet different, sense than were the British. And the Salk controversy was discussed almost immediately on a point-by-point basis in the daily written media. The essential difference as one progressed from the early twentieth century into mid-century and beyond is the speed of the communication.

Those questions provide ample avenues for further research into the relation of the press to technology. And, doubtless, there are other questions which will arise. Whatever the direction, one must concede the following: The press does affect technological change and technological change is reflected in differing ways in the press, reinforcing, of course, that the press offers but one avenue for demonstrating that the culture-technology relation (or technology-culture relation, if preferred) cannot be explained away with unilateral simplicity.